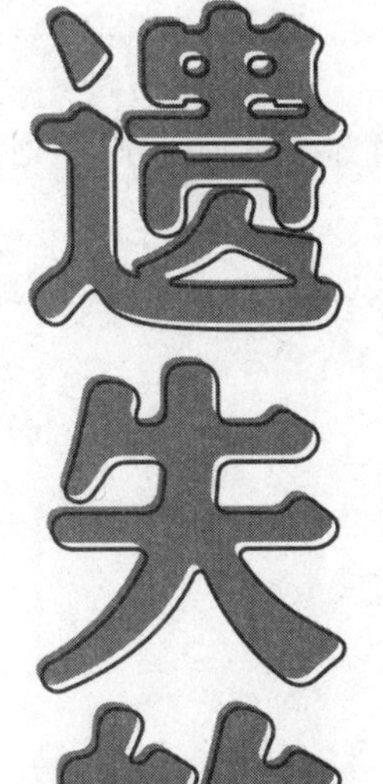

遗失的创意

［英］詹姆斯·摩尔　保罗·尼禄◎著

James Moore & Paul Nero

陈明晖　孙　怡　刘晓燕◎译

陈新忠　王俊英◎注

刘　夙　王　茜◎校订

科学普及出版社

·北　京·

图书在版编目（CIP）数据

遗失的创意. 下册：汉、英 /（英）詹姆斯·摩尔，（英）保罗·尼禄著；陈明晖，孙怡，刘晓燕译. —北京：科学普及出版社，2021.1

书名原文：Pigeon Guided Missiles

ISBN 978-7-110-10011-0

Ⅰ. ①遗… Ⅱ. ①詹… ②保… ③陈… ④孙… ⑤刘… Ⅲ. ①创造发明—青少年读物—汉、英 Ⅳ. ①N19-49

中国版本图书馆 CIP 数据核字（2019）第225668号

CONTENTS
目录

PART ONE 第一部分

M
A
D
H

PART ONE

第一部分

ABANDONED LONDON'S EIFFEL TOWER

注释

① patriotic [ˌpætrɪ'ɒtɪk] *adj.* 爱国的
② lofty ['lɒftɪ] *adj.* 崇高的
③ riposte [rɪ'pɒst] *n.* 机敏的回答
④ inaugurate [ɪn'ɔːgjʊˌreɪt] *v.* 使正式就任
⑤ audacious [ɔː'deɪʃəs] *adj.* 大胆的
⑥ Liberal MP, 全称 Liberal Member of Parliment; The Liberal Party 自由党
⑦ magnate ['mægneɪt] *n.* 巨头，大亨
⑧ edifice ['ɛdɪfɪs] *n.* 大厦
⑨ counterpart ['kaʊntəˌpɑːt] *n.* 对应的人或物
⑩ majestically [mə'dʒestikəli] *adv.* 雄伟地，庄严地，威严地
⑪ quintessential [ˌkwɪntɪ'sɛnʃəl] *adj.* 典型的

When the Eiffel Tower was opened in Paris in 1889 it became the tallest man-made structure on the planet and made the city the envy of the world. Patriotic[①] Briton Sir Edward Watkin thought Victorian London should respond with its own rival tower, vowing: 'Anything Paris can do, we can do bigger.'

Today, few of those who gather at that modern-day temple to lofty[②] aspirations, Wembley Stadium, realise that they are sitting on the very site of Watkin's attempt to pull off his 1,150ft architectural riposte[③] to the French. Yet in the early 1890s, just a few years after the 984ft Parisian landmark was inaugurated[④], it looked as if this audacious[⑤] Liberal MP[⑥] and railway magnate[⑦] might get his way. For a magnificent steel edifice[⑧], planned to be over 150ft taller than its French counterpart[⑨], was beginning to rise majestically[⑩] over the skyline.

Watkin was the quintessential[⑪] nineteenth-century entrepreneur, who believed that modern engineering and technology could overcome almost any obstacle. Aged 72 when work began on the tower in 1891, he was already a

伦敦的埃菲尔铁塔

巴黎的埃菲尔铁塔于1889年正式落成，是当时地球上最高的人工建筑，也使巴黎成为全世界都羡慕的城市。爱国的英国人爱德华·沃特金爵士认为，**维多利亚时代**的伦敦也应该建一座与埃菲尔铁塔相媲美的高塔，他向世人郑重宣告："巴黎能做到的，我们能做得更好。"

时至今日，人们聚集在代表崇高抱负的当代圣殿温布利体育场时，几乎没人意识到他们所处的正是沃特金试图修建351米（1150英尺）高塔以还击法国的地方。19世纪90年代初，那座300米（984英尺）高的巴黎标志性建筑落成短短几年之后，这位勇于创新的**自由党**国会议员和铁路大亨似乎可能如愿以偿，因为一座计划比法国埃菲尔铁塔高46米（150英尺）的宏伟钢结构建筑正准备在天际线处巍然升起。

沃特金是典型的19世纪的企业家，他相信现代工程技术几乎可以战胜一切艰难险阻。1891年铁塔开始

注释

维多利亚时代前接乔治时代，后启爱德华时代，被认为是英国工业革命和大英帝国的鼎盛时期。它的时限常被定义为1851—1901年，即维多利亚女王的统治时期。维多利亚时代与爱德华时代一同被认为是大英帝国的黄金时代。

注释

自由党是英国乃至世界上历史最为悠久的现代政党之一。它曾与保守党在第一次世界大战之前构成英国的两党制。第一次世界大战后自由党的地位逐渐被工党所取代，成为英国的第三大党。

注释

① parliament [ˈpɑːləmənt] *n.* 议会
② flounder [ˈflaʊndə] *v.* 陷入困境
③ thirst [θɜːst] *n.* 口渴
④ multitude [ˈmʌltɪˌtjuːd] *n.* 大量
⑤ stirring [ˈstɜːrɪŋ] *adj.*（活动、演出或讲述）激动人心的

successful man, even by the exacting standards of the industrial age. He had set up a newspaper; helped build railways in nations as far removed as Canada and India; had risen to become director of several railway companies, including Brunel's Great Western Railway; and was a knighted member of parliament①. While his attempt to build the first Channel Tunnel had floundered②, if anyone could build a British Eiffel Tower, he seemed a good man to have behind the project.

The site was to be Wembley, then a village of around 3,000 souls. As befits a true man of money, Watkin chose the location, then in open countryside and some 8 miles from the centre of the city, because it tied in nicely with his other projects. Watkin was already chairman of the nation's first underground line, the Metropolitan Railway, which had opened in the 1860s and in the 1880s was being extended – running north-west out of London. Opening a station at Wembley, and a pleasure park too, would attract people to use the new line as well as satisfy the nation's thirst③ for engineering and industrial prowess. He would also have been aware that the Eiffel Tower was bringing in a multitude④ of visitors and had repaid the cost of its construction in just one year.

Some 280 acres were purchased for the site of Watkin's Tower, sitting at 170ft above sea level. At first, in what was either a gesture of goodwill or breathtaking cheek, Watkin asked Gustave Eiffel himself to design his tower. Unsurprisingly that offer was declined, with Eiffel remarking that the French: 'would not think me so good a Frenchman as I hope I am'. Instead, in November 1889, a competition was launched for new designs for Watkin's tower, stirring⑤ much excitement. The designs poured in – sixty-eight in all – and were exhibited at the hall of Drapers' Company. One was modelled on the Tower of Pisa but without the famous lean; another was to be some 2,000ft

动工时，就算用工业时代严格的标准衡量，72岁的他也已经算是位成功人士了：他创办了一份报纸；帮助各国修建铁路，远至加拿大和印度；担任几家铁路公司的董事长，包括布鲁奈尔大西部铁路公司；他还是拥有爵士爵位的国会成员。虽然他修建第一条英吉利海峡隧道的努力已经付诸东流，但如果说有人能修建英国的埃菲尔铁塔的话，他似乎是完成这一工程的不二人选。

建塔的地址选在了温布利，这里当时是一个约有3000人的村庄。作为一个有经济头脑的人，沃特金选择了这个地方，这里当时属于开阔的乡村，距城镇中心大约12.87千米（8英里），与他的其他工程正好连成一片。沃特金已经是国内第一条地铁——大都会铁路的董事长，这条线路开通于19世纪60年代，在80年代不断延伸至伦敦的西北部。沃特金认为，在温布利修建车站和游乐场不仅会吸引人们来乘坐这条新线路，还会满足国家对工程和工业卓越能力的渴望。他应该还想到了埃菲尔铁塔能吸引大批游客，而且在短短一年之内就能收回建设成本。

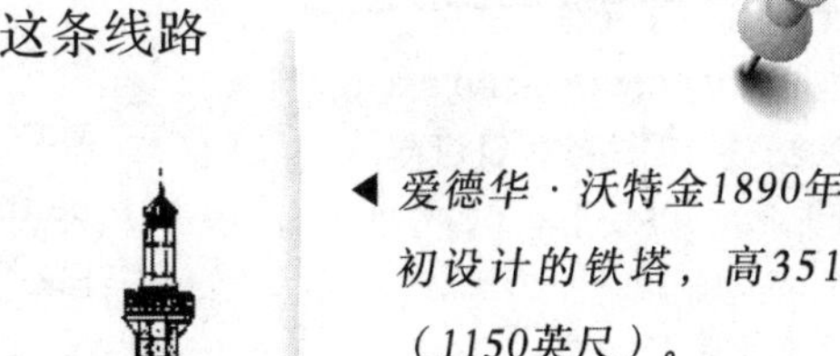

◀ *爱德华 · 沃特金1890年最初设计的铁塔，高351米（1150英尺）。*

Original 1890 design for Sir Edward Watkin's tower, which would have been 1,150ft tall.

沃特金购买了约113公顷（280英亩）土地用于建塔，这块地的海拔为51.82米（170英尺）。起初，沃特金非常善意或极其谦卑地邀请古斯塔夫 · 埃菲尔（Gustave Eiffel）亲自设计这座塔。不出所

high; one featured a spiral① railway which would wind around the tower; and another was to house a 1/12th scale model of the Great Pyramid of Giza; one was even to be called: 'Muniment② of Hieroglyphics③ Emblematical④ of British History During Queen Victoria's Reign Tower'. Alas, the judges decreed⑤ that none of the designs would work as they stood and Sir Benjamin Baker, designer of the Forth Bridge in Scotland, was drafted in to help make one of the designs, which sported eight huge, spider-like, iron legs, more practical.

Watkin formed the Metropolitan Tower Construction Company to oversee the work, which in total was projected to cost £200,000. The tower, due to be completed in 1893, would weigh in at a hefty⑥ 7,000 tons, with lifts transporting visitors to the top. Some 60,000 people were expected per day. Photographs from the time reveal a structure that looks very much like a half-built Eiffel tower. It was, however, to be thinner and with four levels. Visitors would not only get a bird's-eye-view of the capital and surrounding countryside, but the idea was that it would house everything from restaurants to theatres, with exhibitions and even a Turkish baths. At the top there would be an observatory⑦ and the park below was to have a boating lake and a spectacular waterfall.

By 1891 construction on what was now being called The Wembley Tower had begun. But subscriptions⑧ were poor and construction dragged⑨. The press at the time had mixed feelings about Wembley Park. The *Pall Mall Gazette* opined: 'The chief attraction is the Great Tower now in course of construction.' And the *Daily Graphic*, of 14 April 1894, noted that 'over a hundred and fifty men are constantly employed in putting the tower together'. *The Times* talked breathlessly of how an express lift would take impatient visitors to the top in just two and a half minutes. But an article in the *Building News*, also

注释

① spiral ['spaɪərəl] *adj.* 螺旋形的
② muniment ['mjuːnɪmənt] *n.* 防卫手段
③ hieroglyphic [ˌhaɪərə'glɪfɪk] *adj.*（尤指古埃及的）象形文字的
④ emblematical [ˌembli'mætik] *adj.* 象征的，典型的
⑤ decree [dɪ'kriː] *v.* 发布命令
⑥ hefty ['hɛftɪ] *adj.* 庞大的，沉重的
⑦ observatory [əb'zɜːvətərɪ] *n.* 天文台，气象台，瞭望台
⑧ subscription [səb'skrɪpʃən] *n.* 会员费，捐赠款，征订费
⑨ drag [dræg] *v.*（费力地）拖

料，埃菲尔拒绝了，他说："那样我在法国人民心中就没有我自己希望的那样好了。"于是，1889年11月，开展了一场为沃特金的塔做新设计的竞赛，这让许多人兴奋不已。设计方案纷至沓来，共有68个，全部展览在布商公司的大厅里。有一个设计模仿了比萨斜塔，但没有倾斜；有一个设计塔高610米（2000英尺）；有一个设计的特色是螺旋式铁轨绕塔而上；有一个设计是在塔里面放一个大小为原型1/12的埃及吉萨金字塔模型；还有一个甚至把设计的高塔叫作维多利亚时期英国历史的象形标志塔。评审官裁定这些设计都不可行，并选定苏格兰福斯大桥的设计者本杰明·贝克爵士（Sir Benjamin Baker）帮助完善其中一个八脚蜘蛛形钢柱支撑的铁塔，使其更为可行一些。

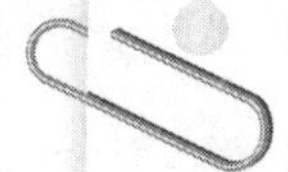

▼ 照片所示为爱德华·沃特金爵士在温布利所建铁塔一期工程完工时的场景，这座塔原计划比巴黎的埃菲尔铁塔还要高。

Photograph showing the first, completed stage of Sir Edward Watkin's Wembley tower. The edifice was planned to be taller than the Eiffel Tower in Paris.

为了监督这项工程，沃特金成立了大都会铁塔建设公司，计划共花费20万英镑。这座塔定在1893年完工，总重预计达7000吨，配有把游客送到塔顶的电梯，预计日游客量6万左右。当时的一张照片上所展示的建筑很像建了一半的埃菲尔铁塔，但比埃菲尔铁塔细一些，而且只有四级。登临铁塔，游客不仅能鸟瞰伦敦和周边郊外的景色，而且塔里设施齐全，有餐厅、剧院、展厅，甚至还有土耳其蒸气浴。塔顶还将有个瞭望台，塔下的公园里有可供泛舟的湖和壮观的瀑布。

published in 1894, called the tower an 'unfinished ugliness'.

Eventually the first level was completed and lifts installed. In 1896 the tower was opened to the public. It had already reached a height of 155ft and was dominating the surrounding landscape. In the same year a local brewer① was using a picture of the tower on advertising to sell their wares②. Then, things began to go drastically wrong. It emerged that the foundations of the tower had begun to move. Poor surveying had failed to identify that the marshy③ ground could not support the weight of the tower with just four legs. Nor had Watkin realised that, unlike the Eiffel Tower, his tourist attraction was at that time too far from the city centre to bring sufficient visitors to pay for its construction, even though his own new railway extension ran right past it. Apart from the station and the parkland, many of the much-vaunted④ facilities had failed to materialise, meaning that in its unfinished state the tower wasn't much of a draw.

Money began to run out, and with only £87,000 raised by 1899 the company went bust and work stopped. There the tower remained, gently rusting⑤, a sorry reminder of the failed venture. In the coming years it was used for army signalling and the odd firework display and dubbed⑥ the London Stump or Watkin's Folly⑦. Finally, on 7 September 1907, it was blown up. The 2,700 tons of scrap metal that had gone into its construction were exported to Italy.

The site's famous days weren't over though. In 1923 Wembley Stadium – with its iconic twin towers - was built on the site. And remains of Watkin's structure were unearthed when the new Wembley Stadium, which opened in 2007, was being built. All that is left of Watkin for posterity is a Watkin Road, a few streets from the current stadium, lined with industrial premises⑧ in stark contrast to⑨ the local 'sylvan⑩ scenery' of the area described by *The Sunday* Times in the 1890s.

注释

① brewer ['bruːə] *n.* 啤酒酿造者，啤酒酿造公司
② wares [wɛəz] *n.* 商品
③ marshy ['mɑːʃɪ] *adj.* 沼泽的，泥泞的
④ vaunt [vɔːnt] *v.* 自夸
⑤ rust [rʌst] *v.* 生锈
⑥ dub [dʌb] *v.* 把……称为
⑦ folly ['fɒlɪ] *n.* 愚蠢的事
⑧ premise ['prɛmɪs] *n.* 经营场所，办公场所
⑨ in stark contrast to 与……形成鲜明的对比
⑩ sylvan ['sɪlvən] *adj.* 树林的，树木的

到1891年，当时所称的温布利塔开始动工，但捐款太少，工期遭到拖延。当时的媒体对温布利公园怀有复杂的感情。《蓓尔美街报》认为："公园的主要景点就是正在施工的高塔。"1894年4月14日的《每日画报》指出："为了建这座塔，陆续雇用了150多人。"《泰晤士报》为直达电梯在两分半的时间内把迫不及待的游客送到塔顶而激动不已。但1894年的《建筑报》中有篇文章把这座塔称为"未完工的丑八怪"。

最终，塔的第一级完工并装上了电梯，于1896年向公众开放。这时塔高已经达到47米（155英尺），在周围也算是鹤立鸡群。同年，当地的一家酿酒商在卖酒广告中使用了塔的图片。接着情况开始直转急下，塔的根基开始动摇。建塔前的地质勘察做得太差，未能发现那儿的湿地无法只靠四根支柱支撑起那么重的塔。沃特金也没有意识到，他的旅游景点位置没有埃菲尔铁塔的好，在那时来说离市中心太远，即使他的铁路已经延伸到那里，但吸引的游客并不算多，无法收回建设成本。除了车站和公园用地之外，很多大肆夸耀的设施都未能建成，意味着未完工的塔无法吸引足够的游客。

资金开始耗尽，到1899年只筹到8.7万英镑，于是公司破产，工程停工。塔留在那里，慢慢生锈，成为这次投资失败的标志物，令人非常伤心。后来，这座塔被用作发放军事信号和表演烟火，被人们戏称为"伦敦树桩"或"沃特金的杰作"。最终这座塔于1907年9月7日被炸毁，建塔所用的2700吨废铁被运往意大利。

然而这个场地继续声名赫赫。1923年，温布利体育

注释

① crumble ['krʌmbəl] *v.* 碎裂，弄碎
② terminus ['tɜːmɪnəs] *n.* 终点站
③ hoard [hɔːd] *n.* 藏品

Watkin himself died in 1901, shortly after it became obvious that his dream had crumbled[①]. Two years before his death, however, he was present at the opening of another more successful venture – Marylebone Station, the terminus[②] for the Great Central's London Extension, one of the last mainline railways to be built in Britain. Despite his north-London folly, his railways – and Wembley Park station – are still in use by millions. To this day hoards[③] of football fans use his line to flock to the spot carrying their own hopes of glory.

场及其标志性的双子塔在这里建成。新的温布利体育场于2007年对外开放，修建时挖出了沃特金所建之塔的残骸。沃特金为后世留下的只有一条沃特金路，它离现在的体育场有几条街远，沿路排列的是工业厂房，而完全不是19世纪90年代《星期日泰晤士报》所描述的“森林风光”。

1901年，沃特金在梦想破灭之后不久就离开了人世。然而，他去世两年前出席了另一次比较成功的投资的开通典礼——马里波恩站，是大中央线延伸到伦敦的终点站，这条铁路线也是英国最后修建的铁路主干线之一。尽管在伦敦北部建了那么一座工程浩大而不适用的塔，但是他的铁路——包括温布利公园站——仍然有上百万人在使用。直到今天，球迷们仍然沿着他铺下的铁路线，纷纷聚集到承载他们胜利希望的足球场中。

POSTPONED

THE FIRST CHANNEL TUNNEL

注释

① burrow ['bʌrəʊ] *v.* 翻找
② interlude ['ɪntəˌluːd] *n.* 间歇
③ trundle ['trʌndəl] *v.*（车辆）缓慢行进
④ ventilation [ˌventɪ'leɪʃ(ə)n] *n.* 通风，换气

Just 21 miles separate Britain from the coast of France at the narrowest point of the English Channel, or La Manche, as the French call it. Here, it is, on average, only around 120ft deep. So building a tunnel underneath this stretch of water was certainly not beyond the Victorians' capabilities. Much of the technological know-how about how to burrow[1] beneath it and link the two nations did indeed exist a hundred years before today's rail tunnel was built. They nearly did it too. In 1880, using state of the art technology, boring began. Prime Minister William Gladstone and his wife even went down. As did the Archbishop of Canterbury. So what happened to this grand scheme and why was it another hundred years before digging began in earnest once more?

As early as 1802 Albert Mathieu-Favier, a mining professor, had put forward the plan of a tunnel during a short interlude[2] in the wars between Britain and France. His idea would have seen horse-drawn carriages trundling[3] through with oil lamps lighting the way. There would be a place to change horses half way through and enormous chimneys to the surface of the water giving ventilation[4]. One cartoon print from the era has the

第一条英吉利海峡隧道

英国从英吉利海峡（法语称拉芒什海峡）最窄处到法国海岸只有33.8千米（21英里）。这里的水深平均大概只有36.58米（120英尺），所以维多利亚时期的英国人自然有能力在这里建一条海底隧道。今天的铁路隧道建成前的一个世纪，事实上已经具备了在水下挖通隧道、连接两国的专业技术，当时的人也差一点就打通了隧道。1880年，采用当时最先进的技术的钻孔开始了，就连首相威廉·格莱斯顿（William Gladstone）和他妻子也深入隧道现场视察，坎特伯雷（Canterbury）大主教也去了。那么这个宏伟计划出现了什么问题？为什么过了一个世纪才再次点燃了人们挖掘隧道的热情？

早在1802年，矿业专家阿尔贝·马提厄-法维耶（Albert Mathieu-Favier）在英法战争短暂的休战期间就提出了修建隧道的计划。他的设想是油灯照路，马车驶过隧道，半路上应该有换马的地方，还应该有

Napoleonic armies triumphantly[1] marching through a tunnel underneath to conquer England with hot air balloons flying across in support.

By the latter half of the century the technological expertise was in place to make the theory of a tunnel a practical reality. Soil surveys showed that the chalk[2] under the Channel would be relatively easy to bore through. Other projects had already proved what could be achieved with modern machines. In 1843 Marc Isambard Brunel's tunnel, 1,300ft under the Thames, had been opened successfully. The 100-mile Suez Canal had opened to shipping in 1869, while an 8-mile railway tunnel had been built through the Alps in 1871 in what many engineers consider tougher geological circumstances. The aftermath[3] of the Franco Prussian war of 1870 brought a tentative rapprochement between France and Britain, with Germany now seen as a common threat, creating a political environment conducive to a tunnel attempt. In 1873 *The Railway News* even ran an article saying that a Channel Tunnel was not only practicable but might pay for itself, too. But it would take private capital to give the plan the impetus[4] it needed.

Step forward Sir Edward Watkin. Some years before his attempt to build a rival to France's Eiffel Tower (see Chapter 11) he was trying to build a railway link to the country too. Watkin, chairman of the South Eastern Railway, thought it a perfect opportunity to fulfil his dream of a railway that would eventually connect his Great Central line from Sheffield to London and right through to Paris. His Submarine Continental Railway Company aimed to tie up with a tunnel being built by a French group on the other side, led by Alexandre Lavalley, contractor for the Suez Canal. A trial tunnel was thus drilled[5] near Dover in 1880. An inscription[6] by one of the workers on the remains can still be seen today. It reads in rather

注释

① triumphant [traɪˈʌmfənt] *adj.*（因胜利或成功而）得意洋洋的
② chalk [tʃɔːk] *n.* 白垩
③ aftermath [ˈɑːftəˌmɑːθ] *n.*（灾难性大事件的）后果
④ impetus [ˈɪmpɪtəs] *n.* 推动力，促进因素
⑤ drill [drɪl] *v.* 钻（孔），打（眼）
⑥ inscription [ɪnˈskrɪpʃən] *n.* 铭文，碑文

巨大的烟囱通到海面换气通风。当时的一幅漫画上画着拿破仑的军队在空中热气球的支援下威武地穿过海底隧道，占领英国。

19世纪后半叶，又出现了新的专业技术，可以使海底隧道从理论变为现实。土壤勘测表明，英吉利海峡之下的白垩层相对容易钻透，其他工程也证明了海底隧道是现代机械可以达到的水平。布律内尔（Marc Isambard Brunel）主持修建的隧道位于泰晤士河下396.24米（1300英尺），在1843年成功开通了。1869年，161.93千米（100英里）的苏伊士运河开通航运。1871年，虽然很多工程师都认为阿尔卑斯山脉的地质很坚硬，但12.87千米（8英里）的铁路隧道穿山而过。1870年普法战争后，英法试图恢复友好关系，德国成为两国的共同威胁，这为修建隧道营造了良好的政治氛围。1873年，《铁路报》甚至发表文章说海底隧道不仅现实可行，而且可能收回成本，但需要私人资本推动计划落实。

爱德华·沃特金爵士（Sir Edward Watkin）站出来了。他曾试图修建一座能与法国的埃菲尔铁塔一争高下的建筑，在这之前他还想修建一条通到法国的铁路。沃特金是东南铁路的董事长，他认为这是个绝佳的机会，能实现他的铁路梦想，最终建成贯通从谢菲尔德到伦敦、继而直达巴黎的大中央线。他的水下大陆铁路公司要与另一端由法国在建的海底隧道合作，法国那端是由**苏伊士运河**的承包商亚历山大·拉瓦雷（Alexandre Lavalley）负责的。因此，1880年，他们

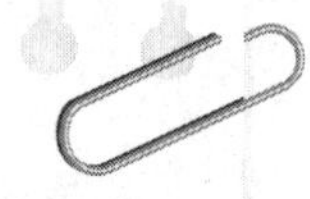
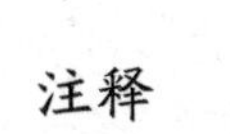

注释

苏伊士运河于1869年修筑通航，沟通地中海与红海，提供从欧洲至印度洋和西太平洋的最近航线。它是世界上使用最频繁的航线之一，也是亚洲与非洲的交界线，是亚洲与非洲、欧洲来往的主要通道。

appropriately faltering[①] English: 'This Tunnel Was Begubugn William Sharp in 1880.' By 1881, the digging[②] of a 7ft tunnel was well under way 100ft below the sea from Shakespeare Cliff, near Folkestone, with good progress being made on the French side, at Sangatte, near Calais.

The speed of the operation was helped by a new wonder – the Beaumont and English boring machine, which had a rotating[③] cutting head to slice[④] through the rock. Within five years, the pilot tunnel connecting the two countries would be finished, announced Watkin. By 1883 a total of 6,178ft had been excavated[⑤] on the English side and 5,476ft on the French side. They were over a mile out from each coast. Watkin predicted 250 trains, powered by compressed air, could be going through the tunnel every day.

But military top dogs, most prominently[⑥] a war hero called General Sir Garnet Wolseley, were getting twitchy[⑦]. He even went so far as warning that legions[⑧] of French troops could come over disguised[⑨] as tourists then suddenly seize Dover. He warned a parliamentary commission into the affair: 'No matter what fortifications[⑩] and defences were built, there would always be the peril[⑪] of some continental army seizing the tunnel exit by surprise.' Watkin lobbied[⑫] hard carrying out 'personally conducted tours and picnics'. But the generals had much of the press on their side, tapping into the public's ancient fear of invasion. The Channel had, many believed, safeguarded the country for hundreds of years. Even the *Railway News* was now dead against the idea.

The government subsequently caved under the pressure and in July 1883 the Board of Trade put a stop to the work. Watkin appealed to common sense, saying that if the government feared a French invasion, they would be able to flood the tunnel, blow it up or simply fill it with smoke and that, if necessary, elaborate

注释

① faltering ['fɔːltərɪŋ] *adj.*（尝试、努力、行动等）犹豫的，蹒跚的
② dig [dɪg] *v.* 挖掘
③ rotating [rəʊ'teɪtɪŋ] *adj.* 旋转的
④ slice [slaɪs] *n.*（切下的食物）薄片
⑤ excavate ['ɛkskəˌveɪt] *v.* 挖掘（古物）
⑥ prominent ['prɒmɪnənt] *adj.* 重要的，著名的
⑦ twitchy ['twɪtʃɪ] *adj.* 焦躁不安的
⑧ legion ['liːdʒən] *n.* 军团
⑨ disguise [dɪs'gaɪz] *v.* 掩饰
⑩ fortification [ˌfɔːtɪfɪ'keɪʃən] *n.* 防御工事
⑪ peril ['pɛrɪl] *n.* 极大危险
⑫ lobby ['lɒbɪ] *v.* 游说

在多佛港附近钻了一条试验隧道，今天仍然能看到遗迹上一名工人刻下的文字，用英语歪歪扭扭地写着："这条隧道是1880年由B. 威廉 · 夏普修建的。"到了1881年，从福克斯顿附近的莎士比亚悬崖开始挖的2.13米（7英尺）隧道正在距海面30.48米（100英尺）的海底顺利施工中，法国在加莱附近的桑加特也进展顺利。

一项新的奇迹加快了工程进度——博蒙特英吉利钻孔机，这种机器带有旋转式切头，能切开岩石。沃特金宣称，试建的这段连接两国的隧道五年之内就能完工。到1883年，英国那边已经挖了1883.05米（6178英尺），法国挖了1669.08米（5476英尺）。两边都超出海岸1.61千米（1英里）多。沃特金预计每天将有250列压缩空气动力火车穿越隧道。

但是军队高层，最主要是战争英雄嘉内特 · 沃尔斯利（Garnet Wolseley）上将变得焦躁不安。他甚至警告说，法国的大部队可以伪装成游客过来然后突然占领多佛。他警告议会委员会介入这件事："无论防御工事多坚固，总会存在欧洲大陆军队突然占领隧道出口的危险。"沃特金通过"导游带领的团队游和野餐"艰难地做着游说工作。但是将军们通过利用公众对侵略的古老恐惧把多数媒体争取到了自己一边。很多人相信，海峡已经保卫了这个国家几百年的安全。就连《铁路报》现在也坚决反对这个工程。

政府后来没能抵挡住压力，1883年7月，贸易工业部叫停了这项工程。沃特金求助于常识，说如果政府害怕法国入侵，可以把隧道淹没、炸掉或者简单点——用烟熏，如果有必要，可以在英国的出口端精心修建防御工事。但没人听他

注释

① plea [pliː] *n.* 恳求
② constraint [kən'streɪnt] *n.* 限制
③ opt [ɒpt] *v.* 选择

fortifications could be built at the British end. His pleas[①] fell on deaf ears. Watkin never quite gave up while he was still alive. Rumour had it that he had even sponsored Gladstone's trip to Paris in 1889 hoping that the tunnel project could be revived. It was to no avail. There was to be no Channel Tunnel out of England for another hundred years. For decades, military opposition and then economic constraints[②] would sink several new ideas and suggestions for a Channel link.

As late as the 1980s, there were plans to forget building a tunnel and build a bridge instead. In fact the £3 billion suspension bridge, which would have been 220ft high, almost got the go ahead from the then prime minister Margaret Thatcher. It would have seen motorists charged just £5.60 to cross between the two countries, taken ten years to build and involved the use of a massive 450,000 tons of steel. However, a Department of Transport memo from 1981 concluded that the bridge would be subject to too many safety and maintenance issues. The government finally opted[③] for a subterranean link and the Channel Tunnel opened in 1994.

的。沃特金在有生之年一直没有完全放弃。有流言说他甚至在1889年赞助了格莱斯顿去巴黎旅行，希望隧道工程能重新开工。但没有起到丝毫作用，接下来的一个世纪英国都不会有海底隧道。几十年来，军方反对加上经济制约使得几次用隧道连接两国的新想法和建议都未能实现。

到了20世纪80年代，政府又制订了修桥而不是修建隧道的计划。事实上，这座将耗资30亿英镑、高达67.06米（220英尺）的吊桥差点得到了当时的首相玛格丽特·撒切尔的批准。这座大桥将用10年时间建成，耗费45万吨钢材，机动车横跨两国只需交费5.6英镑。然而，1981年，交通运输部的一份备忘录认为，跨海大桥存在太多安全问题和维护问题。政府最终选择了地下连接，于1994年开通了**英吉利海峡隧道**。

注释

英吉利海峡隧道于1994年5月6日开通，英吉利海峡隧道位于英国多佛港与法国加来港之间。英吉利海峡隧道由三条长51千米的平行隧洞组成，总长度153千米。

CANCELLED

NELSON'S PYRAMID

注释

① pillar ['pɪlə(r)] *n.* 柱，台柱，顶梁柱

② seafaring ['siːˌfɛərɪŋ] *adj.* 以航海为业的，定期出海旅行的

③ exotic [ɪg'zɒtɪk] *adj.*（常因来自遥远的他国而显得）奇异的

④ replica ['rɛplɪkə] *n.*（雕像、建筑物或武器等的）复制品

⑤ trivial ['trɪvɪəl] *adj.* 无关紧要的

⑥ strip [strɪp] *v.* 剥离

⑦ muscly ['mʌsli] *adj.* 肌肉发达的，强壮的

⑧ nymph [nɪmf] *n.* 宁芙，希腊和罗马神话中的自然女神，常常化身为年轻女子

⑨ realm [rɛlm] *n.*（活动、兴趣、思想的）领域

⑩ eponymous [ɪ'pɒnɪməs] *adj.*（戏剧、书中男女主角）与作品同名的

Trafalgar Square, at the very centre of London, contains what is almost certainly the most famous statue in the country: a pillar①, 170ft high with a 17ft-tall Horatio Nelson on top. But the square might have looked very different if the judges of a competition to design a monument to one of the country's greatest seafaring② heroes had matched his bravery.

The options in front of the Nelson Memorial Committee varied from the absurdly exotic③, such as a replica④ of the Roman Coliseum, to the ridiculously trivial⑤ in the form of a pair of marble dolphins. Often Nelson featured in numerous shapes and sizes; occasionally he was replaced with an attractive alternative, perhaps a seafarer stripped⑥ to a muscly⑦ waist, maybe three nymph⑧-like mermaids. But long before the competition, a plan for the site was put forward by an enterprising, energetic and eccentric Irish knight of the realm⑨, Colonel Sir Frederick William Trench MP. His eye-catching idea: a British pyramid.

At the time that Nelson met his end at the battle of Trafalgar in 1805, his now eponymous⑩ square was known as the King's Mews, housing only unpleasant dwellings for the poor and

纳尔逊的金字塔

在伦敦市中心**特拉法尔加广场**上，矗立着一座可谓全国最有名的雕像——一根高52米（170英尺）的柱子，顶部是5.2米（17英尺）高的霍雷肖·纳尔逊（Horatio Nelson）雕像。当时为了纪念这位英国最伟大的航海英雄，举办了一场设计纪念碑的比赛，如果这次比赛的裁决与这位英雄的英勇事迹相配的话，那么这座广场可能就与现在大不相同了。

摆在纳尔逊纪念委员会面前的竞选方案千奇百怪，有的想法荒诞怪异，譬如模仿**罗马竞技场**；有的想法则平庸至极，譬如一对大理石海豚。通常雕像

注释

特拉法尔加广场，英国伦敦著名广场，坐落在伦敦市中心，是为纪念著名的特拉法尔加海战而修建的，广场中央耸立着英国海军名将纳尔逊的纪念碑和铜像。纳尔逊在这场战役中击溃法国及西班牙组成的联合舰队，使拿破仑彻底放弃海上进攻英国的计划，他自己在战事中中弹身亡。

注释

罗马竞技场是古罗马帝国专供奴隶主、贵族和自由民观看斗兽或奴隶角斗的地方，是古罗马文明的象征，遗址位于意大利首都罗马市中心。

fine horses for the king. And although the recently deceased Nelson was already written into legend, thought didn't turn to a permanent memorial for a generation. In the meantime, Colonel Trench decided that the thrashing[1] of the French, not the beatification[2] of Nelson, was something that should be monumentally marked in the capital.

In 1815, Trench began his campaign with the same interfering style which had been making him a nuisance throughout London for much of his life. Dismissing plans for a new royal palace in Whitehall as 'narrow tradesman-like views', he upset the aristocracy[3], architects and the workers in one deft[4] sentence. Despite this, Trench was, he claimed: 'an advocate for the splendour[5] and magnificence of the crown', and in this respect he was determined to spend the nation's crowns on a dramatic structure, London's largest, in what was to become Trafalgar Square.

Trench thought the pyramid would be the nation's most celebrated building, exceeding[6] the height, the footprint and the glory of St Paul's Cathedral. With each of its twenty-two stepped piers[7] representing a year of the recent wars with France, the structure would remind a grateful nation how Bonaparte's fleet, superior in numbers but lesser in talent, commitment and honour, had been defeated in the Battle of the Nile. After Napoleon set his sights on building an empire along British lines, the Middle East campaign became a turning point in the Anglo-French wars, with the French leader taking Malta on his way to seize Egypt. Napoleon's troops took Cairo easily, but then Nelson's stunning[8] attack at Abu Qir Bay, off the port of Alexandria on 1 August 1798, overwhelmed[9] most of the thirteen French ships. Just four escaped. Balancing Nelson's regret that he had missed a full house, British cannons had happily sliced off the right arm, then the left arm and then one

注释

① thrashing ['θræʃɪŋ] *n.* 轻松击败
② beatification [bɪˌætɪfɪ'keɪʃ(ə)n] *n.*（罗马天主教）宣福
③ aristocracy [ˌærɪ'stɒkrəsɪ] *n.* 贵族阶级
④ deft [dɛft] *adj.* 灵巧的
⑤ splendour ['splɛndə] *n.* 壮美的外观
⑥ exceeding [ɪk'siːdɪŋ] *adj.* 非常的，超越的
⑦ pier [pɪə] *n.* 凸式码头
⑧ stunning ['stʌnɪŋ] *adj.* 令人震惊的
⑨ overwhelm [ˌəʊvə'wɛlm] *v.*（强烈地影响而）使不知所措

被设计为外形各异、大小不一的纳尔逊的形象；有时又被设计为吸引人的其他形象，或是一个上身赤裸、肌肉发达的海员，或是三个仙女一样的美人鱼。但早在比赛之前，出现了一个方案，它的提议者是国会议员及上校弗雷德里克·威廉·特伦奇爵士（Sir Frederick William Trench），这是一位积极进取、充满活力、性情古怪的爱尔兰爵士。他的想法非常吸引眼球——建一座英国金字塔。

1805年，纳尔逊在特拉法尔加海战中壮烈牺牲，那时这座现在以他名字命名的广场名为“国王的马厩”，广场上仅有一些供穷人居住和饲养皇室马匹的破旧住宅。虽然纳尔逊去世不久，他的传奇已被载入史册，但并未有建造世代永久纪念碑的想法。同时，特伦奇上校认为，纪念的对象应当是对法国人的痛击，而不是纳尔逊给英国带来的福音。

1815年，特伦奇开始宣传活动，所用方式和他的想法一样咄咄逼人，这使他后来遭到所有伦敦人的厌恶。他把在白厅修建一座新皇宫的方案贬斥为“狭隘的、充满铜臭味的想法”。仅凭这句妙语，他就同时让贵族、建筑师和工人都非常不爽。不过，如特伦奇所说，他是“皇家荣耀与威严的忠实拥护者”，因此他决心用国家的钱在特拉法尔加广场上修建一座引人注目的、伦敦最大的建筑。

根据特伦奇的设想，这座金字塔将成为整个英国最著名的建筑，比圣保罗大教堂更高、覆盖面积更大、更为壮观。金字塔上有22个阶形墩，每个代表近年来英法交战的一年，以使国人谨记，波拿巴的舰队虽然在数量上占优势，但战术、士气和声望都不及英军，最终法军在尼罗河

leg off one French captain, Aristide Auber Dupetit-Thousars, who nonetheless[1] continued commanding his ship after being placed in a tub of bran[2]. This, to the one-armed, half-blind Nelson, was sweet justice.

Colonel Trench believed this was the Napoleonic battle to commemorate. Trafalgar may have been the decisive[3] battle of Anglo-French wars, but it proved fatal for Nelson. Egypt, by contrast, was a more exotic event in a greatly romantic setting, in which Nelson only suffered a head injury and was later able to report his 'irresistible' pleasure at the discipline of his officers and men. All in all, thought Trench, this was a battle worth spending £1 million to remember. The pyramid's erection was 'an expense not burthensome to the nation' he told incredulous[4] sponsors[5] while asking for the money to cover the cost.

To prove the seriousness of his intentions, Trench commissioned a model from architects Philip and Matthew Cotes Wyatt, sons of James Wyatt the Fonthill Abbey designer (we will meet in Chapter 14). Needing big names to support the plan, he asked the former commander-in-chief of the army, the Duke of York, second son of mad King George III, to provide royal patronage[6] and allow his Pall Mall residence to be opened for an exhibit. The duke[7], not known for sound judgment or strong intellect (he was an incorrigible[8] drunk, womaniser and gambler[9] in the finest traditions of the Hanoverian monarchy[10]), agreed. To the great loss of a nation in need of a pyramid, the exhibition failed to attract the support Trench expected. A busybody, eccentric MP, joining forces with a stupid, drunken duke (it was he who is often credited with marching 10,000 men to the top of the hill and marching them down again) and asking for a cool £1 million for a monument that would dwarf St Paul's, impressed few. Nelson's pyramid got no further than the model in the Duke of York's home.

注释

① nonetheless [ˌnʌnðə'lɛs] *adv.* 然而

② bran [bræn] *n.* 麸

③ decisive [dɪ'saɪsɪv] *adj.* 决定性的

④ incredulous [ɪn'krɛdjʊləs] *adj.* 怀疑的

⑤ sponsor ['spɒnsə] *n.* 赞助机构，倡议人

⑥ patronage ['pætrənɪdʒ] *n.* 赞助

⑦ duke [djuːk] *n.* 公爵

⑧ incorrigible [ɪn'kɒrɪdʒəbəl] *adj.* 屡教不改的

⑨ gambler ['gæmblə] *n.* 赌徒

⑩ monarchy ['mɒnəkɪ] *n.* 君主制，王室

海战中败北。在拿破仑决定沿英国边界建立一个帝国后，这场中东的海战成为英法战争的转折点，那时这位法国领袖在占领埃及的途中拿下了马耳他。1798年8月1日，在拿破仑军队轻松占领开罗后，纳尔逊在亚历山大港阿布吉尔湾出奇兵发起突袭，击败了拿破仑13艘战舰的大部，只有4艘逃脱。虽然纳尔逊很遗憾未能将敌军一举歼灭，但英军火炮将一名名叫阿里斯蒂得·奥伯·迪珀蒂—图萨尔的法军上尉右臂、左臂和一条腿先后炸飞，而这位将领却继续在麦麸桶里指挥作战。对于在战争中丧失了一只眼和一只手臂的纳尔逊而言，这也算扯平了。

▲ 纳尔逊柱所处位置差点竖起的金字塔，由菲利普和马修·科特斯·怀亚特设计，预计耗资100万英镑。

The pyramid which might have been erected where Nelson's column stands today. Designed by Philip and Matthew Cotes Wyatt, it would have cost £1 million.

特伦奇上校认为，这场海战才是最值得纪念的拿破仑战役。虽然特拉法尔加海战也许是英法战争具有决定性的一场战役，但纳尔逊却在这场海战中不幸阵亡。相比之下，尼罗河海战的地点——埃及，则更富异域情调和浪漫色彩。纳尔逊在这次战役中只是头部受伤，而且得到了他后来所说的从约束官兵中得到的一种“无法抗拒”的快感。总之，特伦奇认为这场战役值得花100万英镑来纪念。在募集资金的时候，他对那些心存疑虑的捐助者说，修建这座金字塔“会花一些钱，但绝不会拖累国家”。

为了证明自己是认真的，特伦奇专门委托建筑

But that didn't stop changes to King's Mews. By 1830, horses and houses had been cleared away to leave an open space, the name changing to Trafalgar Square five years later. The whole area around Whitehall by this time resembled① nothing more than a building site: the nearby Houses of Parliament burning down in 1834 just as King's Mews lay fallow. Construction of a lasting Nelson testimonial② remained years away, but at last things were moving. A committee of the great and the good began to organise a competition and to raise funds for a statue, dolphin, Coliseum or whatever it was going to be, although certainly not a pyramid.

Respondents were invited to think freely, but when it came, the choice of a dull but functional and unquestionably tall column outraged③ almost as many people as had gasped④ at Trench's pyramid. However, Nelson's column it was to be, despite strong protests that dogged it throughout construction between 1840 and 1843. A disproportionately⑤ small Nelson, a tenth the size of the column he would stand on, was a folly that the public would not take to its heart, some said. No one would be able to see the great hero, said others, with not a little justification. It would spoil the view of Whitehall

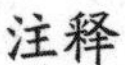

注释

① resemble [rɪ'zɛmbəl] *v.* 像
② testimonial [ˌtɛstɪ'məʊnɪəl] *n.* 纪念品，奖品
③ outrage ['aʊtreɪdʒ] *v.* 使震怒
④ gasp [gɑːsp] *v.* 倒吸气
⑤ disproportionate [ˌdɪsprə'pɔːʃənɪt] *adj.* 不成比例的

师菲利普（Philip）和马修·科特斯·怀亚特（Matthew Cotes Wyatt）制作了一个金字塔模型。这两位建筑师都是詹姆斯·怀亚特（James Wyatt）（放山修道院设计师）的儿子。鉴于这个计划需要大人物的支持，特伦奇邀请前军队统帅、约克公爵、疯狂国王乔治三世的次子给予皇家赞助，并开放他的帕尔摩住宅展览这个模型。这个在世人看来糊里糊涂的公爵（他也是一个继承了汉诺威君主制优良传统的无可救药的酒鬼、登徒子和赌徒）同意了他的请求。可是，令特伦奇失望的是，展览并未得到预期的支持。如果不能修建金字塔，这对国家可是一个巨大损失。确实，一个古里古怪、多事的议员和一个愚蠢的、醉醺醺的公爵（据信他曾率领1万人行军至山顶，然后又行军下山），要向人们募集整整100万英镑修建一座比圣保罗大教堂更宏伟的纪念碑，没有多少人能相信他们。纳尔逊金字塔的最终结果也只是约克公爵家里的那个模型。

然而这也没有挡住“国王的马厩”发生变化。1830年，那里马去房空，五年后又改名为特拉法尔加广场。此时，白厅四周就像是一个建筑工地：就在“国王的马厩”还是一片空地的时候，附近的议会大厦已在1834年被烧毁，而修建纳尔逊的永久纪念碑不知要等到猴年马月了。不过事情终于有了进展，一个由一些大人物组成的委员会开始组织一场比赛挑选方案，筹资修建一座雕像。海豚也好，竞技场也好，反正这次不会再是金字塔了。

委员会邀请许多人自由思考，但当结果出来时，这个毫无创意、仅具象征性、明显过高的纪念柱一公布就引起了公愤，绝不亚于当年特伦奇的金字塔方案在公众中引起

注释

① bashing ['bæʃɪŋ] *n.* 严厉批评，猛烈抨击
② miserable ['mɪzərəbəl] *adj.* 痛苦的
③ compulsory [kəm'pʌlsərɪ] *adj.* 强制性的

and parliament from the north. Even a parliamentary select committee joined the column bashing①, but its hands were tied: 'We do not see how Government can avoid interfering to prevent the miserable② result of which they are so clearly forewarned ... Alas for the prospects of the Art in England.'

Today, London has three notable monuments to this period in British history. Nelson's column, but more broadly Trafalgar Square has, after all, won its place it the nation's affections. And the grand old Duke of York, the man who housed the model of Nelson's pyramid, has a 123ft-tall column of his own, paid for by a compulsory③ tax on British soldiers, each of whom had to give up a day's wages to fund it when he died. Frederick Augustus Hanover, Duke of York, is on the Duke of York Column in Waterloo Place, a quarter of a mile away from Nelson. And on the Embankment rests Cleopatra's Needle, a gift from the Egyptians to the British in 1819 that honours the very same battle that Trench wanted to remember: Nelson's Battle of the Nile.

的震惊。然而，最终还是确定了纳尔逊柱方案，尽管直到1840—1843年施工期间，这个方案仍然遭到了强烈反对。一些人说，纳尔逊雕像高度只有他脚下柱子高度的1/10，比例太小，看上去十分愚蠢，公众是不会把这样一个雕像放在心上的；另一些人说，没有人能看得到这位伟大英雄，因为他太小了，这听起来也有道理。它还会阻碍北面的白厅和议会的视线。连议会特别委员会也加入了抨击这根柱子的队伍，但也只能说："我们无法理解政府是如何回避这么多反对意见的，早就应该向他们提醒这样一个可悲的结果了……唉，想想英国艺术的未来吧！"

今天，伦敦有三座纪念这段英国历史的纪念碑。纳尔逊纪念柱，范围再大点就是特拉法尔加广场，终究在国民心中赢得一席之地。那个家藏纳尔逊金字塔模型的约克公爵，也为自己立了一根高37.5米（123英尺）的纪念柱。纪念柱是以强制税的形式筹资修建的：在他去世后每个英国士兵上交他们一天的工资作为强制税。约克公爵弗雷德里克·奥古斯·汉诺威的纪念柱，现位于滑铁卢广场，距纳尔逊纪念柱0.4千米（0.25英里）。此外，泰晤士河河堤上还矗立着克利奥帕特拉之针（古埃及方尖碑）。这是1819年埃及送给英国的礼物，为纪念特伦奇想让人们记住的同一场战役——纳尔逊的尼罗河海战。

CANCELLED

WREN'S MISSING MARVELS

In September 1666, as diarist Samuel Pepys wept on the south bank of the Thames watching London burn on the opposite bank, an opportunist astronomer hatched① an extravagant② plan. From the ashes of the city – a city that, up until that point, had been a higgledy-piggledy③ mess of urban depravity④ and privilege in equal measure – he would build a new capital; the most majestic⑤ capital city in Europe. Christopher Wren's qualifications for such a task may have been lacking – at the time, his most significant brush with architecture was upsetting and almost coming to a fist⑥ fight with the designer improving St Paul's Cathedral – but he made up for his deficiencies⑦ with enthusiasm, intelligence and force of personality.

Recently returned from an extended tour of Paris, a catastrophe⑧ of biblical proportions presented Wren with a never-to-be-repeated opportunity: the chance to create a new city on a blank, and rather charred⑨, canvas. To this point,

注释

① hatch [hætʃ] *v.* 使孵出，孵出

② extravagant [ɪk'strævəgənt] *adj.* 奢侈的，浪费的

③ higgledy-piggledy ['hɪgəldɪ'pɪgəldɪ] *adj.* 乱七八糟的，杂乱无章的

④ depravity [dɪ'prævɪtɪ] *n.* 堕落

⑤ majestic [mə'dʒɛstɪk] *adj.* 壮丽的，雄伟的

⑥ fist [fɪst] *n.* 拳头

⑦ deficiency [dɪ'fɪʃənsɪ] *n.* 缺乏，不足

⑧ catastrophe [kə'tæstrəfɪ] *n.* 灾难

⑨ charred [tʃɑ:d] *adj.* 烧焦的

雷恩错失的奇迹

1666年9月，当记载了当时情景的日记作者塞缪尔·佩皮斯（Samuel Pepys）走上泰晤士河南岸，看着对岸的伦敦大火哀悼时，机会主义者、天文学家克里斯托弗·雷恩（Christopher Wren）酝酿了一个大胆的计划：他打算在这座城市的废墟上重建新首都，打造全欧洲最宏伟的首都。当时的伦敦，权贵林立、乌烟瘴气。要完成这么艰巨的任务，雷恩的资历或许有些不足。那时，他最重要的建筑作品充满了争议，他甚至差点儿与改造圣保罗大教堂的设计师打了一架，但是他的激情、才智与人格魅力弥补了自身的缺陷。

刚从巴黎长期旅行回国后不久，雷恩就目睹了一场巨大灾难，这给他带来了一生难遇的机会：在一块空白的烧焦画布上创造一座新城市。那时，雷恩的事业虽然比较成功，但仍只限于学术，而非实践领域。他30多岁就已经是牛津大学的天文学教授，已经达到了英国学术界的巅峰。他爱好写诗，写得最好的一首二十行华丽韵文是关于一位

注释

① studious ['stju:diəs] *adj.* 好学的，用功的
② pinnacle ['pɪnəkəl] *n.* 顶峰
③ dabble ['dæbəl] *v.* 涉猎
④ florid ['flɒrɪd] *adj.* 过分花哨的
⑤ resuscitation [rɪˌsʌsɪ'teɪʃən] *n.* 复苏，复兴
⑥ graphite ['græfaɪt] *n.* 石墨
⑦ lubricate ['lu:brɪˌkeɪt] *v.* 使润滑
⑧ spleen [spli:n] *n.* 怨气，怒气
⑨ spaniel ['spænjəl] *n.* 西班牙猎犬
⑩ emanate ['ɛməˌneɪt] *v.* 发散出（品质）
⑪ scramble ['skræmbəl] *v.* 争夺
⑫ reliant [rɪ'laɪənt] *adj.* 依赖的，依靠的
⑬ urine ['jʊərɪn] *n.* 尿
⑭ parmesan [ˌpɑ:mi'zæn] *n.* 帕尔马干酪（一种浓味硬奶酪，常用于意大利饮食中）

Wren's career, though distinguished, was studious[1] but not practical. Still in his 30s, as professor of astronomy at Oxford University, he had reached the pinnacle[2] of English academia. He'd dabbled[4] in poetry; his best attempt being twenty lines of florid[4] verse about the resuscitation[5] of a hanged woman. And he had produced a series of relatively unimpressive inventions: a device for writing in the dark; some graphite[6] for lubricating[7] timepieces; and a demonstration model of how a horse's eye worked. His experiments with animals – including removing the spleen[8] of his spaniel[9] to see how it would get on without one – had become tiresome. It was time for a career change.

So Wren, and London, had a lot to thank Thomas Farryner for. At about 1 a.m. on Sunday 2 September, Farryner, a baker who supplied the navy with ships' biscuit, noticed a smell emanating[10] from elsewhere in his home in Pudding Lane. It was the Great Fire of London, and it was burning in the room next door. Waking his family, Farryner managed to scramble[11] to safety, leaving only a housemaid, too frightened to flee, to the mercy of the flames and to become the fire's first victim.

London's fire risk had long been evident. Tinderbox wooden homes, lit by candles, heated by open fires; buildings packed tightly together; with fire-fighting reliant[12] on buckets of sand, milk and, less convincingly, urine[13]. After two days of chaotic attempts to extinguish the blaze, the situation looked worse, not better. On Tuesday, St Paul's caught light, causing Pepys, who had been following the fire's progress in his diaries, to feel the situation serious enough to bury his papers, his wine and a rather splendid parmesan[14] cheese he had been working through. By Wednesday, with the blaze finally out, about 200,000 people were homeless. More than 13,200 houses, a hospital, the Royal Exchange, three city gates and, to the delight of inmates, Newgate Prison had been destroyed: 80 per cent of the city

受到绞刑的女子复活的故事。此外，他还有一些不太知名的发明：用于在黑暗中写字的装置；润滑钟表的石墨；演示马的眼睛运作机理的模型。他对动物实验（包括切除他的西班牙猎犬的脾脏，看其能否正常生活）也日渐生厌。真该换个职业了！

所以，雷恩和伦敦都应该感谢托马斯·法里纳（Thomas Farryner）。法里纳是为海军军舰供应面包的面包师。1666年9月2日（星期日）凌晨1点左右，法里纳闻到布丁巷的家里散发出一种气味，原来是伦敦着火了，隔壁的房间也着火了。法里纳叫醒家人，逃往安全的地方。但是一个女佣由于胆子太小，不敢冲出火焰，成了这场大火的第一个遇难者。

伦敦早就存在火灾隐患：易燃的木屋、照明的蜡烛、取暖的明火和密密麻麻的房子，而火灾发生时，灭火仅靠一桶桶的沙子、牛奶和尿，真是令人难以置信。经过两天乱哄哄的救火之后，情况似乎变得更糟了。到了星期二，圣保罗大教堂也着火了。佩皮斯一直在日记中记录火灾的进展情况，他看到这种情形，感觉不妙，就把日记本、酒和一块相当美味的帕尔马干酪埋了起来。直到星期三，火才熄灭了，大约20万人无家可归。大火烧毁了13200多间民房、1家医院、皇家交易所、3座城门，还有新门监狱（犯人们高兴坏了），伦敦城80%的地方化为灰烬。虽然大火只直接造成9人丧生，但由于人民因此深陷犯罪、饥荒和贫困之中，真正的伤亡人数要多得多。

当人们还在激烈争论失火原因时（他们认为666是个邪恶数字，是1666年发生的这场大火的罪魁祸首），雷

reduced to embers[①]. Although only nine deaths were directly attributed to the fire, the true loss is believed to be many more as people fell victim to crime, hunger and poverty.

As debate about the cause raged – 666 being the number of the beast, in 1666 sin was the most obvious culprit[②] – Wren progressed with his plans. Within days, he was in front of the king, Charles II, presenting a breathtaking design for a new London based on classical Paris and Rome. He had been wide-eyed with awe at the architecture in the French capital. In contrast to London's chaotic and still largely medieval streets, Paris, though smaller, was blessed with expansive boulevards[③], pleasant parks and elegant public spaces. Through his friend and competitor John Evelyn, with whom Wren almost certainly shared ideas before or even during the fire, he knew too of 1580s Rome under Pope Sixtus V, with its Palladian buildings, fine avenues and wondrous piazzas.

Wren's new London borrowed from both cities. The main boulevard was to run from Fleet Street to Aldgate, with a circus at the centre. A new Royal Exchange, rebuilt on its original site, would dominate a piazza where ten streets intersected[④]. All the city's main commercial buildings – the Mint, Post Office, goldsmiths, the Excise Office, the banks – formed a commercial district. A second large avenue, with two piazzas along the way, started at the Tower and converged at St Paul's. The cathedral itself would be smaller than its predecessor and the number of churches elsewhere cut from eighty-six to fifty-one. Even the streets proclaimed sense and order, with a grid[⑤] system off the boulevards banishing forever the cramped[⑥], unhygienic[⑦] conditions of London before the fire. In his biography *Parentalia*, Wren's son, also Christopher, claimed the new streets had 'three Magnitudes' – thoroughfares[⑧] at least 90ft wide, secondary streets 60ft, and lanes 30ft. Wren's scale drawing doesn't back up that claim, but the roads were certainly

注释

① ember ['ɛmbə] *n.* 余烬

② culprit ['kʌlprɪt] *n.*（造成问题或麻烦的）原因

③ boulevard ['buːlvɑː] *n.* 林荫大道

④ intersect [ˌɪntə'sɛkt] *v.* 相交

⑤ grid [grɪd] *n.* 网格，（地图上的）坐标方格

⑥ cramped [kræmpt] *adj.* 狭促的

⑦ unhygienic [ˌʌnhaɪ'dʒiːnɪk] *adj.* 不卫生的

⑧ thoroughfare ['θʌrəˌfɛə] *n.* 大道，主要大街

恩却在推进自己的计划。几天后，他站在国王查理二世面前，展示了他根据巴黎和罗马的建筑所设计的新伦敦，让人惊叹不已。他对法国巴黎的建筑充满敬畏。与乱糟糟、大部分仍是中世纪街道的伦敦相比，巴黎面积虽小，却拥有宽敞的林荫大道、宜人的公园和高雅的公共场所。伦敦大火之前甚至大火期间，通过约翰·伊夫林（John Evelyn，雷恩的朋友和竞争对手），雷恩了解了16世纪80年代教皇西斯都五世统治的罗马，那儿有帕拉第奥式建筑、美丽的大街和令人惊叹的广场。

雷恩设计的新伦敦借鉴了巴黎和罗马两座城市的设计。主大街从佛里特街延伸至奥尔德盖德，中心有一个圆形广场。皇家交易所将在原址重建，俯瞰着一个十条街在此交会的广场。城市所有主要商业大厦，包括造币厂、邮局、珠宝商行、税务署、银行等构成商业区。第二大街从伦敦塔至圣保罗大教堂，途经两个广场。而大教堂将比它的前身小，城市中教堂数量也由86个减少至51个。就连街道也彰显了理性与秩序：大道以外的街道组成方格系统，改变了大火之前伦敦街道狭窄、不卫生的状况。雷恩的儿子（也叫克里斯托弗）在父亲的传记《先人传》中说，新街道分“三个等级”：大道至少宽27米（90英尺），次要街道宽18米（60英尺），小巷宽9米（30英尺）。虽然雷恩的比例图并未说明这一点，但马路规模之大为英格兰前所未有。一个崭新的时代即将到来。

雷恩和伊夫林的方案虽然是各自设计的，但却有许多共同之处。雷恩的方案略胜一筹，但有两个致命缺陷。首先，他的设计缺乏精确性，对于即将成为英国历史上最

注释

① eminent ['ɛmɪnənt] *adj.* 卓越的，有名望的
② sequestrate [sɪ'kwɛstreɪt] *v.* 扣押，接收
③ skint [skɪnt] *adj.* 身无分文的
④ veto ['viːtəʊ] *v.* 否决
⑤ faff [fæf] *v.* 小题大做，无事瞎忙
⑥ sip [sɪp] *v.* 小口地喝

of a scale unseen in England before. A new era was coming.

Wren's and Evelyn's plans, though developed separately, shared much in common. Wren's was the finer, but it contained two fatal flaws. Firstly, it was based on inaccurate plans of the area, an elementary mistake for someone who was to go on to be the most eminent[①] architect in British history. Secondly, and much more difficult to solve, it would take a very long time to build. Wishing to keep the people on side, the king wanted normality returned quickly. Wren's scheme, if it was presented with any seriousness at all – and that, from an astronomer, was in doubt – would involve undue delay. Moreover, Wren hadn't taken account of existing ownership rights. People would have to be compensated for having their land sequestrated[②]. And the city was skint[③].

Just days after both Wren and Evelyn presented their plans, Charles vetoed[④] them both as too complicated and costly. Most of London's income came from property, and most of that property was in ruins. The only way of getting the city back on its feet was to allow people to rebuild their houses and return to commerce, not faff[⑤] about sipping[⑥] coffee on a boulevard like they did in Paris.

Other architects rushed to Whitehall with schemes of their own, including Wren's friend Robert Hooke, and one Captain Valentine Knight, whose over-enthusiastic suggestion that the king might be on to a nice little earner from the rebuilding resulted in a spell in prison. Not one of the plans was implemented. From the safe distance of half a century, Wren's son attacked the king for not possessing sufficient foresight to understand the transformative benefits. Wren's own transformation from astronomer to architect was nevertheless under way. Along with Evelyn, Hooke and three other eminent men, the king appointed Wren to a commission which was to be funded, in the most British of ironies, with a new tax on coal –

杰出建筑师的他来说，这是一个低级错误；其次，工程耗时过长，这个问题就更难解决了。为了安抚民心，国王想尽快恢复正常。雷恩的方案过于耗时，加之这个方案出自一位天文学家之手，人们难免怀疑。同时，雷恩没有考虑土地的现有产权。政府要征收老百姓的土地就必须补偿他们，而那时的伦敦已经不名一文了。

雷恩和伊夫林提交方案后没几天，查理二世就把两个方案都否决了，理由是二者都太复杂、太费钱了。当时伦敦的主要收入来自房地产，而大部分房产已经毁掉了。重振这座城市的唯一办法就是让老百姓重建住所、恢复商业，而不是像巴黎人那样坐在林荫大道上闲聊和品咖啡。

另外一些建筑师也带着各自的方案涌入白厅，他们中有雷恩的朋友罗伯特·胡克（Robert Hooke），还有一位名叫瓦伦丁·奈特（Valentine Knight）的上尉，他热情洋溢地建议国王从伦敦重建中赚一小笔，这个主意导致他在牢里待了一段时间。这些方案没有一个被采纳。半个世纪之后，议论国王也不会再有什么危险了，雷恩的儿子批评国王，认为他缺乏远见，不知道重建伦敦能带来哪些好处。不过雷恩的职业生涯却已经开始从天文学家转向建筑师了。国王任命雷恩、伊夫林、胡克及另外三位杰出的建筑师成立了一个委员会。该委员会由一项新的煤炭税来提供资金——对煤炭征税的理由是多数人使用煤炭，而且偷带者很难让自己身上不沾上炭黑。这恐怕是最具英国特色的幽默了吧！

六名委员认为，伦敦仍然可以拥有壮观的新林荫大道和建筑，虽然会多花点钱和时间。但当这个问题被拿到议

注释

① sooty ['sʊtɪ] *adj.* 沾满烟灰的
② parish ['pærɪʃ] *n.* 教区
③ muster ['mʌstə] *v.* 聚集（支持、力量、精力等）
④ gild [gɪld] *v.* 镀金
⑤ urn [ɜːn] *n.* 骨灰瓮
⑥ engraving [ɪn'greɪvɪŋ] *n.* 版画

the rationale being that most people used it and it was hard to smuggle without the offender ending up sooty[①].

The six commissioners argued that London could still have splendid new boulevards and buildings, despite the cost and delay. But while the king suggested a middle way – a new quay and wider streets – when parliament debated the issue they quickly got bored and decided their time would be better spent working out how to escalate a conflict with Holland. (The Dutch were rumoured to be arsonists who had started the Great Fire in the first place in revenge for the English burning one of their harbour towns just weeks before.) All building plans were scaled back, the work was divided out and eventually Wren – later Sir Christopher – went on to design all fifty-one new parish[②] churches in the city, together with his new landmark, St Paul's Cathedral. This, at £265,000, cost just over a third of the money raised by the coal tax. A tribute to the Great Fire designed by Wren and Hooke was erected near the place where the Great Fire ignited. But even this was not the monument that Wren first suggested. He originally wanted a phoenix on top of a column, but he decided that people wouldn't be able to identify the bird 200ft high above them. A statue of Charles in Roman costume didn't pass muster[③] with the king who thought it would be too expensive, and why didn't Wren think of something along the lines of a large ball of metal that people could see at a distance? So the monument today is a Doric column with a flaming gilded[④] urn[⑤].

The story does have a partially happy ending, however, for Wren's new London does exist in part in one capital city. Working from his original engraving[⑥], in 1799, US President Thomas Jefferson laid out what they then called Federal City. Today, Wren's London grid forms the basis of a part of Washington DC.

会上讨论时，国王提出了一个折中的建议——修建一个新码头和更宽的街道。议员们很快就厌烦了，他们认为应当把精力放在如何升级与荷兰的战争上（传说荷兰人就是这起大火的始作俑者，他们这么做是为了报复几周前英国人放火烧了他们的一个港口镇）。这样，所有建筑计划都被缩减了，工作也被分配了下去。最终雷恩（后来的克里斯托弗爵士）继续负责设计城区的51座教堂和他里程碑似的新作——圣保罗大教堂。这些建筑共花费了265000英镑，只占煤炭税收的1/3多一点。雷恩和胡克还设计了伦敦大火的纪念碑，把它安放在大火起源地附近。但这座纪念碑与雷恩的初衷不同，他最初希望纪念柱顶上是一只火鸟，后来又觉得人们无法看清距离头顶61米（200英尺）高处的鸟，所以就放弃了。另一个方案是身着罗马风格服饰的查理二世雕像，因为太昂贵不符合国王的要求也作罢了。为什么雷恩就没想到类似于老远就能看到的大金属球这样的设计思路呢？于是今天的纪念碑就是一根顶部为火焰镀金骨灰坛的多立克柱。

尽管如此，故事的结局还是有可喜可贺之处，因为雷恩的新伦敦设计确实有一部分存在于某个首都城市。1799年，美国总统托马斯·杰斐逊（Thomas Jefferson）采用雷恩的最初模型，设计了他们当时所称的“联邦城”。今天，雷恩的伦敦格局形成了华盛顿特区一部分地区的基础。

DISASTER

THE TUMBLING ABBEY HABIT

He was 'England's wealthiest son', a man of such means he could build a cathedral[①] to live in if he wanted and get the greatest architect of his generation to design it. This combination of unlimited money and unrivalled[②] talent should have resulted in a secular abbey so well constructed that it would survive the centuries; so beautiful that it would compete for the nation's affections with Salisbury or Winchester or York. Yet William Beckford's abbey at Fonthill, Wiltshire, which spanned[③] 270ft from east to west and 312ft from north to south, with a central tower reaching almost 400ft to its turrets[④], stood for just a generation.

The whole enterprise should be a lesson to all homeowners who, thinking they know better than their professionals, overrule them on crucial issues, change their minds after the plans are agreed, and who bribe[⑤] builders with liquor. For Beckford did all this and more. Spoilt[⑥], eccentric, yet not unkind or unintelligent – he had an important art collection and was the author of bestselling gothic[⑦] novel *Vathek* – his intention was to build an impenetrable structure where he could shut himself away from a hostile world and live out his days in peace.

注释

① cathedral [kə'θiːdrəl] *n.* 大教堂，主教座堂

② unrivalled [ʌn'raɪvəld] *adj.* 无与伦比的

③ span [spæn] *v.* 持续

④ turret ['tʌrɪt] *n.* 角楼，塔楼

⑤ bribe [braɪb] *v.* 贿赂

⑥ spoil [spɔɪl] *v.* 毁坏，破坏

⑦ gothic ['gɔθik] *adj.*（以神秘、怪诞为特征的）哥特派文学的

一场灾难

放山修道院的倒塌

他是“全英格兰最富有的人”，富到可以建一个大教堂供自己居住，并聘请当时最伟大的建筑师来设计。无尽的钱财、无双的才智，本可以打造一座能历经几百年的世俗修道院，可与索尔兹伯里、温彻斯特或约克的修道院媲美，赢得国人爱戴。然而，威廉·贝克福德（William Beckford）位于威尔特郡的放山修道院，东西跨越82.3米（270英尺），南北跨越95米（312英尺），中部塔楼至其角楼高达121.9米（400英尺），却只存在了一代人的时间。

整个工程对于私房屋主来说是一个教训，特别是那些认为自己比专家更懂行、在关键问题上不听从他们的建议、计划定好了又临时改变、还用酒来贿赂建筑商的房主。贝克福德就是这样一位房主，他甚至做得更过火。贝克福德是一个骄奢古怪的人，却也天性良善聪明，他收集了很多重要的艺术品，还是畅销哥特小说《瓦提克》的作者。他想建造一座密不透风的城堡，把自己与这个充满敌意的世界隔离开来，在里面安度晚年。

注释

① reclusive [rɪ'kluːsɪv] *adj.* 独处的，隐居的
② aristocratic [ˌærɪstə'krætɪk] *adj.* 贵族的
③ summons ['sʌmənz] *n.* 传票
④ ludicrous ['luːdɪkrəs] *adj.* 荒谬的
⑤ precocious [prɪ'kəʊʃəs] *adj.* 早熟的
⑥ indiscretion [ˌɪndɪ'skrɛʃən] *n.* 言行失检
⑦ ostracise ['ɒstrəˌsaɪz] *v.* 排斥
⑧ allegation [ˌælɪ'geɪʃən] *n.* 指控
⑨ scurrilous ['skʌrɪləs] *adj.*（指控或故事）恶语诽谤的
⑩ oddball ['ɒdˌbɔːl] *n.* 怪人
⑪ valet ['vælɪt] *n.* 贴身男仆

Beckford had reason to be reclusive[①]. His aristocratic[②] ancestry meant he moved, when the muse very occasionally took him, in important circles. Blood of Edward I, Edward III, the House of Lancaster and the House of Stuart ran through his veins and the family was extremely well connected. When young William showed an interest in music, his father summonsed[③] Mozart to England to provide lessons. 'He was eight years old and I was six,' Beckford recalled. 'It was rather ludicrous[④] one child being pupil of another.' But the chemistry was good; so good that Mozart used bars by Beckford in the *Marriage of Figaro* – at least according to Beckford. But just four years after Mozart's lessons, Beckford's father was dead, leaving William an estate of £1 million. It was a sum – the equivalent of about £200 million in 2011 – that in 1769 could go a very long way indeed.

So precocious[⑤], talented, influential, landed and now just about the wealthiest person in the country, William Beckford had everything going for him. Then an indiscretion[⑥] at the age of 23 (he had fallen in love with the 11-year-old heir to the Earl of Devon) led to him being ostracised[⑦] from society and fleeing the country. This landmark event in his life arguably formed the origins of Fonthill Abbey. Given his high profile, and despite the allegations[⑧] never being proven in court, Britain's newspapers, ever eager for a scurrilous[⑨] story, went to town. Beckford fled to the continent with his adoring wife Lady Margaret Gordon, where she subsequently died in childbirth. This tragedy, combined with the earlier scandal, turned an already eccentric oddball[⑩] into a recluse. With his wife dead, time on his hands and money in his pocket, Beckford decided to travel through Europe, taking along his doctor, baker, cook, valet[⑪], three footmen and twenty-four musicians. He wasn't an easy guest or boss. Hotel managers were required to redecorate

贝克福德的遁世不是没有原因的。当缪斯女神偶尔光顾他的时候，他的贵族血统能使他在充满大人物的圈子里活动。他与爱德华一世、爱德华三世、兰开斯特家族、斯图亚特家族都有血缘关系，而他的家庭也建立了极其广泛的社交。当年轻的威廉显示出对音乐的兴趣时，他父亲就把莫扎特召往英格兰，担任他的音乐私教。“那年他8岁，我6岁。”贝克福德回忆道，“一个孩子教另一个孩子，这真可笑。”但两人很投缘，甚至连莫扎特的作品《费加罗的婚礼》中有几小节就使用了贝克福德创作的旋律——至少贝克福德是这样说的。但就在莫扎特担任威廉音乐教师四年后，威廉的父亲就去世了，给他留下了100万英镑的遗产。这笔财产非常可观——相当于2011年的2亿英镑——在1769年足够用上很长一段时间的。

▲ 放山修道院——威廉·贝克福德巨大的哥特式宅邸，只存在了30年时间。

A view of Fonthill Abbey, William Beckford's enormous gothic home which stood for just thirty years.

早慧、聪明、有权有势，现在又是全国最富有的人，威廉·贝克福德可以说要风得风、要雨得雨。然而，他23岁那年的一次鲁莽举动（他爱上了德文伯爵11岁的继承人）导致他遭受放逐，逃离国外。这件大事可以说是放山修道院的根源。由于他的身份地位，虽然在法庭上对他的指控没有证据，但喜欢散播流言蜚语的英国报纸已将他的丑闻传遍全国。贝克福德与他的爱妻玛格丽特·戈登夫人逃到欧洲大陆，随后

before he arrived, and he expected exceptional service, however impractical: even if it meant shipping a flock of sheep from England to Portugal to improve the view from his window.

After thirteen years of this nomadic① existence, Beckford decided sufficient time had elapsed② for him to return safely to Britain. Back at Fonthill, he tore down his boyhood home, itself a Palladian③ mansion erected by his father only relatively recently, and constructed a very high and extraordinarily long wall behind which plans for the abbey began. He was, according to the *Country Literary Chronicle*④,'determined to produce an edifice uncommon in design, and adorn it with splendour'. With privacy uppermost⑤ in his mind, Beckford was anxious that Fonthill Abbey should be constructed in secrecy, although the sudden appearance of a 6 mile wall – 'the barrier' as it became known – had generated considerable local interest. A small village sprang up to house workers and the area buzzed⑥ with excitement. Villagers knew something big was happening and the secrecy added to the intrigue⑦.

Former surveyor to Westminster Abbey James Wyatt, King George III's favourite architect, was delighted to win the commission for a project of such magnitude⑧ and, not to put too

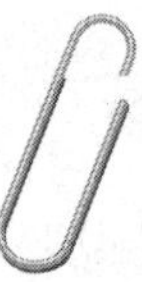

注释

① nomadic [nəʊ'mædɪk] *adj.* 游牧的
② elapse [ɪ'læps] *v.*（时间）流逝
③ palladian [pə'leɪdɪən] *adj.* 帕拉第奥新古典主义建筑风格的
④ chronicle ['krɒnɪkəl] *n.* 编年史
⑤ uppermost ['ʌpəˌməʊst] *adj.* 最高的
⑥ buzz [bʌz] *v.* 嗡嗡地响
⑦ intrigue [ɪn'tri:g] *n.* 阴谋
⑧ magnitude ['mægnɪtju:d] *n.* 量级，巨大

她在那里死于生产。经历了这场悲剧，还有早先的那件丑闻，贝克福德这个本已古怪的人变成了一个遁世者。妻子不在了，他只剩下大把的时间和大把的钞票，于是决定带上他的医生、糕点师、厨师、贴身男仆、3个侍从和24个乐师，游历整个欧洲。贝克福德是一个难对付的客人和老板。在他到达一个酒店之前，酒店经理都被要求重新装修酒店。他还期待额外的服务，无论有多么不切实际：为了让窗外多一道风景，他甚至会要求从英格兰运一群绵羊到葡萄牙。

流浪了13年后，贝克福德觉得是安全返回英国的时候了。回到放山，他把他少年时代住的房子拆了，那是他父亲建的一座帕拉第奥风格的宅邸，还相当新。然后他修了一堵又高又长的围墙，在围墙背后开始建修道院。根据《乡村文学纪事》，贝克福德“决心打造一座设计独特、华丽壮观的宫殿”。由于不想让外界知道这件事，贝克福德极力隐秘地建造放山修道院。可是一道10千米（6英里）长的围墙——人们称之为“堡垒墙”——平地而起，这已经激起当地人极大的好奇心了。一个小村庄突然出现，用来给工人们提供住宿，这个地方也变得热闹起来。村民意识到肯定有什么重大的事情在发生，而工程的隐秘更使人们觉得神秘莫测。

前威斯敏斯特教堂检查员詹姆斯·怀亚特（James Wyatt）——国王乔治三世最宠幸的建筑师，欣然赢得委任建造这样大规模的一个工程，预算非常高。可是建筑师怀亚特太忙了，让他来负责这项工程必然导致后来放山修道院的倒塌。事实上，怀亚特没能给予他手头所有进行中

fine a point on it, with such an obscene[1] budget. The decision to hire Wyatt, an architect who was much in demand, in itself had consequences for the subsequent fall of Fonthill. Wyatt was unable to give all his ongoing projects the attention they deserved and, in the opinion of Prime Minister Lord Liverpool, was an 'incurable absentee; certainly one of the worst public servants I recollect'. More tellingly, Wyatt had earned the nickname 'The Destroyer' after less than exemplary[2] work at Durham, Hereford, Lichfield and Salisbury cathedrals, a slander that Beckford was happy to overlook, for without doubt, Wyatt's plans for Fonthill were exemplary. Towers, spires[3], cavernous[4] rooms, enormous windows capturing splendid vistas[5], Fonthill was perfect – while it stood. It was the workmanship, not the design, which turned out to be shoddy[6]. When Beckford tried to cut corners, Wyatt, for whatever reason, but presumably both absent and exasperated[7] with a contrary client, let him have his way.

The more the builders worked, the more Beckford rushed them. Soon 500 men were employed on the construction, and when the nights began to draw in, they were ordered to work round the clock, by torch and lamplight. He almost doubled numbers by enticing[8] builders working on Windsor Castle, one of Wyatt's projects for George III, with the promise of beer.

Even then, construction, thought Beckford, was stalling. That Europe's grandest abbeys had taken centuries to build was of no consequence. The nineteenth century was upon them and Lord Nelson was coming for dinner. Importing granite[9] or marble or other fine material would waste time. Compo cement[10] and lime[11] mortar[12] that could be rendered[13] with sand to look like stone was ordered instead, with the considerable downside that this substance wouldn't hang together sufficiently to hold parts of the abbey upright. Boozy builders, an impossible timescale

注释

① obscene [əb'siːn] *adj.* 淫秽下流的
② exemplary [ɪg'zɛmplərɪ] *adj.* 堪称典范的
③ spire [spaɪə] *n.*（教堂等建筑物的）尖顶
④ cavernous ['kævənəs] *adj.* 巨穴般的，（房间或建筑物内）如洞穴般空旷的
⑤ vista ['vɪstə] *n.*（尤指从高处看到的）景色
⑥ shoddy ['ʃɒdɪ] *adj.* 粗制滥造的
⑦ exasperated [ɪg'zɑːspəreɪtɪd] *adj.* 恼怒的
⑧ enticing [ɪn'taɪsɪŋ] *adj.* 诱人的
⑨ granite ['grænɪt] *n.* 花岗岩
⑩ compo cement 混合水泥
⑪ lime [laɪm] *n.* 石灰
⑫ mortar ['mɔːtə] *n.* 灰浆
⑬ render ['rɛndə] *v.* 使成为，使变得

的工程应有的关注。在首相利物浦勋爵看来，他是一个“无可救药的缺席者，是我所能记起的最糟糕的公仆”。更值得一提的是，在并不出色地完成了达勒姆、赫里福德、利奇菲尔德和索尔兹伯里大教堂的修复工程后，怀亚特还得了一个“破坏者”的绰号。可贝克福德不顾这些传言，因为怀亚特给出的放山修道院的修建方案显然还是像模像样的。塔楼、尖顶、洞穴般的房间、可以看到壮观景色的巨大窗户……图纸里的放山修道院是完美的。其实问题出在施工上，而不是设计。当贝克福德想走捷径时，无论什么原因，也许是因为不在场，且被这样一个和他唱反调的客户惹怒了，最终怀亚特还是随他去了。

工人们做得越多，贝克福德越是催他们。很快工地上就有500个工人在施工。每当夜幕降临，他们还得凭借火把和灯光继续干活。他还用啤酒把温莎城堡——乔治三世的一个工程——工地上的建筑工人吸引到他的工地上，他的工人人数由此增加了一倍。

即使这样，贝克福德仍认为工程停滞不前。他忽略了一个事实，那就是欧洲最宏伟的修道院是花了几个世纪修建的。可是19世纪就要到来，纳尔逊勋爵即将前来赴宴。由于进口花岗岩、大理石或其他好材料会浪费时间，所以他开始订购当时新研制的一种混合水泥和石灰砂浆作为替代品，它们和沙子混合在一起可以做出石头的效果。这种材料有个严重缺点，那就是要支撑部分城堡，它的黏力还不够强大。醉醺醺的建筑工人、完不成的时间表、劣质的材料，这些都无法建造出坚固的城堡。没多久，一部分建筑就开始崩塌。第一个楼塔修到91米（300英尺）高的时

and inadequate material rarely produce robust[1] results. It wasn't long before parts of the building began to crumble[2]. The first tower, having reached 300ft in height, collapsed initially during a gale[3]. When Beckford ordered it to be rebuilt using the same material, it collapsed again. It was only on the third rebuilding that Beckford agreed to switch to stone.

But once finished, Fonthill Abbey was everything Beckford had dreamed of: magnificent and private. With a much-reduced[4] retinue[5] of servants, including a dwarf[6] whose job description included standing by the 38ft high doors to make them look even taller, Beckford continued to write and collect art. A select number of old friends were invited to inspect the building. 'Who but a man of extraordinary genius would have thought of rearing in the desert such a structure as this, or creating such an oasis?' wrote Henry Venn Lansdown, not realising that by mounting the circular staircase around the towers to admire the view towards the Bristol Channel, he was taking his life in his hands. Nelson eventually visited with Lady Hamilton in 1800, and found himself in the unusual predicament[7] of being publicly criticised for enjoying the hospitality of a man who had once been accused of lewd[8] conduct.

Then another personal disaster struck Beckford. Poverty – at least in relative terms. The collapse of the sugar market had hit his Jamaican plantations, where 1,200 slaves worked the land. Suddenly he needed cash. Fonthill went on the market and the eventual buyer, a gunpowder dealer, was happy with his acquisition, until masonry[9] began to tumble around him. In all, Fonthill Abbey's towers collapsed at least six times. The Great Octagon took its final fall on 21 December 1825 while a subsequent owner, John Farquhar, was elsewhere in the abbey. It was never to be rebuilt. By the 1840s, much of Fonthill was gone forever, the building left in 'desolate[10] ruin'. After the

注释

① robust [rəʊ'bʌst] *adj.* 强有力的（观点、见解）
② crumble ['krʌmbəl] *v.* 崩塌
③ gale [geɪl] *n.* 大风
④ much-reduced [mʌtʃ rɪ'dju:st] *adj.* 大大减少
⑤ retinue ['rɛtɪˌnju:] *n.* 随行人员，随从
⑥ dwarf [dwɔ:f] *n.* 侏儒
⑦ predicament [prɪ'dɪkəmənt] *n.* 困境
⑧ lewd [lu:d] *adj.* 淫秽的，猥亵的，下流的
⑨ masonry ['meɪsənrɪ] *n.* 砖石建筑
⑩ desolate ['dɛsəlɪt] *adj.* 荒凉的

候，就被一阵大风吹垮了。当贝克福德下令用同样材料来重建时，它又塌了一次。直到第三次重建，贝克福德才同意使用石头。

尽管如此，完工后的放山修道院的确符合贝克福德的梦想：宏伟而私密。在新的城堡里，贝克福德减少了侍从数量。侍从中还有一个侏儒，他的一项工作就是站在11.6米（38英尺）高的门口，以使这些门看起来更高。贝克福德则继续写作和收集艺术品。他请了一些老朋友来参观这个建筑。亨利·维恩·兰斯道恩（Henry Venn Lansdown）这样写道："除了一个具有非凡天分的人，还有谁会想到在戈壁中建起这样一座城堡，创造出这样一片绿洲呢？"可他并不知道，当他爬上塔楼的环形楼梯观看布里斯托尔海峡景色时，是在冒着生命危险的。1800年，纳尔逊终于携汉密尔顿夫人前来拜访，可事后纳尔逊发现自己陷于公众舆论中，批评他竟然接受一个曾因猥亵行为而受到指控的人的款待。

接下来，贝克福德遭遇了他的另一场个人灾难——贫穷，至少是相对意义上的贫穷。糖市的崩溃影响了他在牙买加的种植园，那里有1200个奴隶。他突然之间没钱了。放山修道院被拿到市场上拍卖，最终被一个火药商买去了。火药商对这座城堡很满意，直到这个石楼开始在他身边倒塌。算起来，放山修道院的楼塔至少倒塌了六次。1825年12月21日，八角大楼最后一次倒塌。那时，城堡后来的一位房主约翰·法夸尔（John Farquhar）正在城堡内别处。此后，放山修道院再也没有得到重建。到19世纪40年代，放山修道院的许多地方已经永久消失，只留下一片

注释

① secular ['sɛkjʊlə] *adj.* 非宗教的

② adjoining [ə'dʒɔɪnɪŋ] *adj.* 毗邻的

③ sarcophagus [sɑː'kɒfəgəs] *n.* 石棺

death of a further owner in 1858, most of its remains were demolished. Today, part of the north wing which escaped the fall of the tower, and which has its own 76ft-tall turret intact, survives. But Fonthill Abbey, in all its magnificence, had stood for just thirty years. It was one of the shortest-surviving full-sized secular① cathedrals in history. William Beckford himself retired to Bath where he bought two adjoining② properties – connecting them with a new bridge and building a tower on the hill at the end of the road, which is now a museum in his honour. He is buried in the graveyard there, in a pink sarcophagus③ next to his dog.

荒芜的废墟。到1858年下一位房主去世后，大部分残余建筑已被拆毁。今天，只有北侧边房没有受到塔楼倒塌的影响，23米（76英尺）高的角楼仍矗立在上面，完好无缺。但是，放山修道院的辉煌仅持续了30年。它是史上寿命最短的一座完整的世俗大教堂。贝克福德本人则隐退到巴斯，在那里他购置了两个毗邻的房产，建了一座新桥连接两座宅第，还在路尽头的山上修了一个塔楼，现在那个地方已成为以他的名字命名的博物馆。他被埋葬在那的墓地里，他粉色的石棺边还埋葬着他的狗。

CANCELLED

WHY LUTYENS' CATHEDRAL VANISHED

注释

① mourn [mɔːn] *v.* 悼念
② austerity [ɒ'stɛrɪtɪ] *n.* 经济紧缩
③ boast [bəʊst] *v.* 吹嘘
④ crypt [krɪpt] *n.* 教堂地下室
⑤ beguiling [bɪ'gaɪlɪŋ] *adj.* 迷人的
⑥ intricate ['ɪntrɪkɪt] *adj.* 复杂精细的
⑦ diameter [daɪ'æmɪtə] *n.* 直径

The Second World War saw the destruction of many magnificent buildings. In Britain, the loss of Coventry's fourteenth-century cathedral to bombing, is still mourned①. But it wasn't the only great cathedral to fall foul of the conflict. In fact, thanks to the Nazis and the resulting age of austerity②, what might have been the country's most impressive place of worship was never built at all.

Liverpool's lost Roman Catholic cathedral would have boasted③ the world's biggest church dome – a breathtaking edifice which would surely have been one of the architectural wonders of modern times. Today the only part of the structure which was actually built, the crypt④, lies beneath the more humble, modernist, Liverpool Metropolitan Cathedral. Another beguiling⑤ glimpse of what could have been there is still possible, however, thanks to the careful restoration of an amazing 12ft-scale wooden model of the original concept, now on view at the new Museum of Liverpool. It reveals, in intricate⑥ detail, just how ambitious and awe inspiring it was designed to be. The cathedral's spectacular 510ft-high dome, measuring 168ft in diameter⑦, would have been bigger than that

为什么鲁琴斯的大教堂消失了

项目取消

第二次世界大战中，许多宏伟的建筑遭到摧毁。英国人至今仍对考文垂14世纪修建的大教堂被炸毁而深感痛惜，但这座教堂并非唯一一座受到战争破坏的大教堂。事实上，由于纳粹统治及其带来的经济紧缩，一座可能成为英国历史上最伟大的教堂最终未能建成。

这座未建成的利物浦天主教大教堂本来可以拥有全世界最大的教堂穹顶——绝对称得上是现代建筑奇迹的一件令人惊叹的作品。今天这座建筑唯一建成的部分——教堂的地下室，位于相比之下更简朴、更具现代主义风格的利物浦大都会大教堂下面。不过，幸亏人们精心修复了根据最初构想制作的一个令人惊叹的3.7米（12英尺）大的木制模型，我们才有机会一睹这座建筑的迷人风采。这个模型现在陈列于新利物浦博物馆内，上面美轮美奂的细节向世人展示了当初的设计是多么野心勃勃、令人敬畏。根据最初的设计，这座大教堂的穹顶高为156米（510英尺），直径为51米（168英尺），极其壮观，比罗马圣彼得大教

of St Peter's in Rome. At more than 600ft long and 400ft wide, the building would have taken up a massive 6 acres, twice the area as St Paul's in London, radically transforming the city's skyline. Constructed from brick and silver–grey granite, it would have been entered through a huge, soaring arch①.

The person behind this daring blueprint was one of Britain's greatest architects – Sir Edwin Lutyens. Sir Edwin first made his name designing Arts and Crafts-style country houses. He went on to create the Viceroy's Residence in New Delhi and many of the finest First World War memorials, including the Cenotaph in London's Whitehall and the Memorial to the Missing in Thiepval, France. At the height of his reputation in 1929 Lutyens, an Anglican, received what was to be his grandest commission – a Roman Catholic cathedral for Liverpool, set to be the crowning② glory of his career. The city already had a new Anglican③ cathedral designed by Giles Gilbert Scott, himself a Roman Catholic, which was well on its way to being completed. But Liverpool's large Roman Catholic population, which had surged④ to 250,000 in the decades following the Irish potato famine⑤ in the mid-nineteenth century, was crammed into tiny churches, without a defining focal point of worship.

The first attempt to build a Catholic cathedral in the city had failed back in Victorian times, leaving only a Lady Chapel, later demolished⑥. In 1928 the new archbishop⑦, Richard Downey, was determined that Liverpool should get its second cathedral. He sought out Lutyens, the most famous British architect of his day and, according to Lutyens' autobiography, the pair arranged the deal to build it over cocktails! As the project was launched in 1929 Downey announced:

> Hitherto all cathedrals have been dedicated to saints. I hope this one will be dedicated to Christ himself with

注释

① arch [ɑːtʃ] *n.* 拱形结构
② crowning ['kraʊnɪŋ] *adj.* 最伟大的
③ Anglican ['æŋglɪkən] *adj.* 英国国教会的，圣公会的，安立甘会的
④ surge [sɜːdʒ] *v.* 剧增
⑤ famine ['fæmɪn] *n.* 饥荒
⑥ demolish [dɪ'mɒlɪʃ] *v.* 彻底摧毁
⑦ archbishop ['ɑːtʃ'bɪʃəp] *n.* 大主教

堂的穹顶还要大。教堂长逾183米（600英尺），宽逾122米（400英尺），占地2.4公顷（6英亩），是伦敦圣保罗大教堂的2倍，如果建成，将彻底改变整个城市的高空建筑轮廓线。大教堂由砖和银灰色花岗岩建成，入口处是一座极高的拱门。

▼ *鲁琴斯设计的利物浦大教堂的模型，保存于利物浦国家保护中心，目前陈列在新建的利物浦博物馆。*

A model of Lutyens' Liverpool cathedral that never was, restored at National Conservation Centre, Liverpool, and now on display at the new Museum of Liverpool.

这个大胆蓝图背后的设计者是英国最伟大的建筑师之一埃德温·鲁琴斯爵士（Sir Edwin Lutyens）。埃德温爵士最早以设计乡村住宅而出名，后来他又设计了新德里的总督府和多座纪念第一次世界大战的精美纪念碑，包括伦敦白厅的和平纪念碑和法国的蒂普瓦尔失踪者纪念碑。1929年，鲁琴斯的声望正如日中天，这个英国国教教徒接受了他一生中最伟大的使命——为利物浦建造一座天主教大教堂，这将成为他职业生涯的最高荣耀。当时，利物浦已经有一座新建的英国国教大教堂即将完工，设计者是贾莱斯·吉尔伯特·斯科特（Giles Gilbert Scott），他本人是个天主教教徒。尽管如此，利物浦的天主教教徒非常多，在19世纪中期爱尔兰马铃薯饥荒之后几十年内猛增至25万人。这么多天主教教徒要挤到那些小教堂里，很难正常进行神圣的礼拜仪式。

注释

① sanctuary ['sæŋktjʊərɪ] *n.* 避难所
② triumphal [traɪ'ʌmfəl] *adj.* 庆祝胜利的，庆祝成功的
③ synthesis ['sɪnθɪsɪs] *n.* 结合体
④ belfry ['bɛlfrɪ] *n.* 钟楼，钟塔
⑤ vault [vɔːlt] *n.* 拱顶，穹隆
⑥ narthex ['nɑːθɛks] *n.*（教堂的）前廊
⑦ destitute ['dɛstɪˌtjuːt] *adj.* 赤贫的
⑧ unveil [ʌn'veɪl] *v.* 为……揭幕

a great figure surmounted on the cathedral visible for many a mile out at sea.

We do not want something Gothic … The time has gone by when the Church should be content with a weak imitation of medieval architecture. Our own age is worthy of interpretation right now and there could be no finer place than a great seaport like Liverpool.

Brownlow Hill, the site identified for the project, had once housed a huge Victorian workhouse. Now, bought for £10,000, it would be the setting for a place of sanctuary[①].

The scale of Lutyens' grand design was remarkable. Architectural historian Sir John Summerson summed it up as: 'the supreme attempt to embrace Rome, Byzantium, the Romanesque and the Renaissance in one triumphal[②] and triumphant synthesis[③]'. Estimated to cost £3 million, it would be 180ft higher than the neighbouring Anglican cathedral. Along with its dome, the complex structure was to be topped with belfry[④] towers, spires and scores of statues. Inside, granite-lined nave and aisles would feature a series of barrel vaults[⑤] running at right angles, and there was to be a total of fifty-three altars. Inside the West Porch a huge space called the 'narthex[⑥]' would offer a place for the 'cold and destitute[⑦]' to shelter. Even the church's organ was to be the world's largest.

Like the architects of most great medieval cathedrals, Sir Edwin knew that he probably wouldn't see the cathedral completed, believing it would take decades to finish. Downey however was in a hurry – and talked of a 'cathedral in our time'. Lutyens unveiled[⑧] drawings of the cathedral to the public at the Royal Academy in 1932 and his grand £5,000 model of the cathedral was begun. Shortly afterwards Pope Pius XI gave the plans his blessing, and on 5 June 1933 (the year Adolf Hitler

利物浦第一次尝试修建天主教大教堂是在维多利亚时代，不过未能成功，只留下一座圣母堂，后来也被拆毁了。1928年，新上任的大主教理查德·唐尼（Richard Downey）决定在利物浦修建该市的第二座大教堂。他找到鲁琴斯这位当时英国最著名的建筑师，而且鲁琴斯在自传里说，他俩居然喝着鸡尾酒就把修建大教堂这事定了下来！当1929年这个项目开工的时候，唐尼向世人宣布：

> 迄今为止，所有大教堂都是为圣徒而建的。我希望把这座大教堂献给基督本人，在教堂顶上竖立一个海上数英里之外也能看到的巨幅耶稣雕像。
>
> 我们不想要哥特式建筑……教堂对中世纪建筑亦步亦趋的时代已经一去不返了。我们生活的时代值得用建筑来诠释，而选择在利物浦这个伟大的海港修建这座大教堂再合适不过了。

最终选定布朗洛山为大教堂的修建地点，那里曾有一座大型维多利亚济贫院。现在这块地被用1万英镑买了下来，用于建造一座圣殿。

鲁琴斯的设计规模宏大，不同凡响。建筑史学家约翰·萨莫森爵士（Sir John Summerson）将其概括为“融罗马、拜占庭风格、罗马式建筑和文艺复兴时期建筑特点为一体的伟大尝试”。这座大教堂预计耗资300万英镑，比邻近的英国国教大教堂高55米（180英尺）。除了穹顶以外，这个复杂的建筑顶部建有钟塔、尖顶和许多雕像。

was elected in Germany) the first foundation stone was laid at an open-air mass. Work began on the foundations and by the outbreak of war in 1939, an incredible 70,000 tons of earth has been excavated① and 40,000 cubic feet of granite had been laid. The crypt was almost finished. Only half of the growing £1 million fund, mostly raised from parishioners②, had so far been used.

In 1941, however, construction stopped. Instead of a gracious cathedral, barrage③ balloons loomed over Merseyside; the area suffered seventy-nine bombing raids④ during the war with a death toll of 4,000 people. The crypt was used as an air-raid shelter. By the time the war was over, Lutyens had died from cancer.

When the project was finally revived in the early 1950s, costs had risen to a massive £27 million. Archbishop Downey died in 1953 and the new archbishop, Dr William Godfrey, aimed to scale back the original plans. In another twist to the story Adrian Gilbert Scott, brother of the Anglican cathedral's architect, was hired to come up with a design that kept the iconic dome but was much smaller and would only cost £4 million. But the resulting design was still deemed⑤ too pricey⑥. In 1960 a competition was announced for an entirely new cathedral, which would cost a frugal⑦ £1 million. Lutyens' vision was to be finally consigned to history, and in 1967, Sir Frederick Gibberd's circular, concrete cathedral, since dubbed⑧ Paddy's Wigwam, was finished. The then archbishop admitted: 'You can loathe⑨ it, or you can love it – but you can't ignore it.' In the years that have followed the structure has been beset by leaks and corrosion⑩.

The stunning model of Lutyens' cathedral is a stirring⑪ remnant⑫ of the church many would have preferred to see. For Sir Edwin's son, Robert Lutyens, the very fact that his

注释

① excavate [ˈɛkskəˌveɪt] *v.* 挖掘（古物）
② parishioner [pəˈrɪʃənə] *n.* 教区居民
③ barrage [ˈbærɑːʒ] *n.* 连续炮击
④ raid [reɪd] *n.* 突袭
⑤ deem [diːm] *v.* 认为，相信
⑥ pricey [ˈpraɪsɪ] *adj.* 价格高的
⑦ frugal [ˈfruːgəl] *adj.* 俭朴的
⑧ dub [dʌb] *v.* 把……称为
⑨ loathe [ləʊð] *v.* 厌恶
⑩ corrosion [kəˈrəʊʒən] *n.* 腐蚀
⑪ stirring [ˈstɜːrɪŋ] *adj.*（活动、演出或讲述）激动人心的
⑫ remnant [ˈrɛmnənt] *n.* 残余部分，残迹

教堂中殿和走廊以花岗岩镶边，由一连串相互垂直的筒形拱顶组成，教堂中将修建53个圣坛。西门廊内有一块被称为“前厅”的巨大空地，可以为穷人遮风挡雨。就连教堂的风琴也是当时世界上最大的。

和大部分中世纪大教堂的建筑师一样，埃德温爵士知道他可能这辈子也看不到教堂完工，因为他相信这项工程要耗时几十年。不过唐尼却很着急，口口声声说要建“我们这个时代的大教堂”。1932年，鲁琴斯在皇家艺术院向公众揭晓了大教堂的草图，而他价值5000英镑的大教堂模型也开始动工。不久之后，教皇庇护十一世（Pope Pius Ⅺ）同意了这个计划。1933年6月5日（阿道夫·希特勒被选为德国国家元首的那一年），在露天弥撒中举行了奠基仪式，于是地基工程开始了。到1939年战争爆发时，已经挖出7万吨泥土，铺下1133立方米（4万立方英尺）花岗岩，真是令人难以置信。教堂的地下室差不多竣工了。建造教堂的资金主要来自教徒的捐款，总共筹集了100万英镑，而且还在不断增加。到那时，也只用了一半的资金。

然而到1941年，工程停止了。除了阻塞气球隐约出现在默西塞德郡上空，人们再也看不到这座优雅的大教堂升起了。战争期间，这一地区遭受了79次空袭，造成4000人死亡。教堂的地下室成了防空洞。到战争结束时，鲁琴斯已患癌症去世。

20世纪50年代初，这个工程再次启动，造价则攀升至2700万英镑。1953年，大主教唐尼去世，新上任的大主教威廉·戈德弗雷（William Godfrey）博士决定压缩原计划。故事的另一个转折点是阿德里安·吉尔伯特·斯科特

father's lost Liverpool landmark remains only a dream is part of its appeal. He believed it had been saved from 'prejudiced[①] denigration[②]' in its unfinished glory. In a 1969 interview he explained: 'It is there, yet it is nowhere. It is architecture asserted[③] once and for ever – the very greatest building that was never built!'

注释

① prejudiced ['prɛdʒʊdɪst] *adj.* 有偏见的

② denigration [ˌdeni'greiʃən] *n.* 诋毁

③ assert [ə'sɜːt] *v.* 坚定地陈述

（Adrian Gilbert Scott）负责重新设计这座大教堂，他是英国国教大教堂建筑师的弟弟。他的任务是缩小大教堂标志性的穹顶，造价只有400万英镑，但他的设计仍被认为太昂贵。1960年开始了一场修建一个全新大教堂的竞标，造价仅为100万英镑。鲁琴斯的构想最终成为历史，1967年，由弗雷德里克·吉伯德爵士（Sir Frederick Gibberd）负责设计的绰号为“爱尔兰佬的棚屋”的混凝土圆形大教堂竣工了。就连当时的大主教也承认：“你可以厌恶它，也可以热爱它，但你不能忽略它。”在接下来的时间里，这座建筑一直受到漏雨和腐蚀问题的困扰。

鲁琴斯精美绝伦的大教堂模型成为这座教堂留下的遗迹，许多人都想一睹为快。对埃德温爵士的儿子罗伯特·鲁琴斯（Robert Lutyens）而言，他父亲设计的利物浦标志性建筑的梦想未能实现，这才是吸引人的地方。他相信，这个未完成的荣耀使其免于“充满偏见的诋毁”。在1969年的一次采访中，他解释说：“它在那里，可哪儿都找不到它。这是一座从未建成的最伟大的建筑，从一开始就注定了，永远都是！”

ABANDONED

NEW YORK'S DOOMED DOME

With inventions that crossed into the realms① of science fiction, including plans for a Spaceship Earth, 4D towers and a fishshaped car that was steered by a rudder②, Richard Buckminster Fuller was one of the twentieth century's most intriguing inventors and a man with an attuned③ social conscience.

Although many of his ideas never saw the light of day, the 4D concept – referring to how the building would, in addition to its physical three-dimensional properties, be used over time – is a fundamental part of urban design today. This fourth dimension can improve energy efficiency and environmental impact, and, at its best, alleviate④ poverty and inequality. More improbably, his original 1920s' prefabricated⑤ tower blocks were designed to be flown into position across the US by airship. That didn't happen, but the lessons learned during writing the patent application for them would eventually be used in the development of 'geodesic⑥ domes' – large golf ball-like structures that are today widely used for science and entertainment venues⑦. One dome in particular was set for big things: it would cover a huge part of central Manhattan, be energy efficient and have its own

注释

① realm [rɛlm] *n.*（活动、兴趣、思想的）领域
② rudder ['rʌdə] *n.*（飞机的）方向舵
③ attuned [ə'tju:nd] *adj.* 能听出（某种声音）的
④ alleviate [ə'li:vɪˌeɪt] *v.* 减轻（不适）
⑤ prefabricated [pri:'fæbrɪkeɪtɪd] *adj.*（建筑物）预制的，组装的
⑥ geodesic [ˌdʒi:əʊ'dɛsɪk] *adj.* 曲面几何学的
⑦ venue ['vɛnju:] *n.* 举办场所

项目废弃

注定失败的纽约穹顶

理查德·巴克敏斯特·富勒（Richard Buckminster Fuller）是20世纪最引人注目的发明家，具有强烈的社会责任感，他的发明穿越现实，进入了科幻小说之境，包括地球飞船计划、四维塔和具有方向舵的鱼形汽车等。

他的许多想法从未实现，但他的四维建筑概念（指除建筑的三维物理属性外，如何随时间变化应用建筑）却成了今天城市设计的基本考虑因素。第四维可以改善建筑物的能源效率和环境影响，在最佳状况下，还能缓解贫穷和不公平。更令人匪夷所思的是，在20世纪20年代他还曾设想用飞船把预制的摩天大楼运送到美国各地。这个设想最终未能实现，但他在撰写这项专利申请书期间所学到的知识后来在开发“网格穹顶”时派上了用场。网格穹顶是一种高尔夫球状的大型建筑，今天被广泛用于各种科学和娱乐场馆。他设计的一个穹顶非常巨大：覆盖曼哈顿中心的大部分地区，高效节能，自成气候。这个穹顶直径3.2千米（2英里），中心处高1.6千米（1英里），覆盖曼哈顿50个

climate. Two miles in diameter and one mile high at its centre, the domed city would cover fifty of Manhattan's central blocks and benefit from its own weather, which meant no long, snow-clogged winters, and rain only when it was needed.

These were big ideas from a man with an enormous intellectual capacity – he went on to become president of Mensa, the international organisation for people with high IQs. There was personal motivation, steeped in tragedy, too, behind Buckminster Fuller's drive to make the world a better place. At just 4 years of age, his daughter had lost her life to an illness caused by poor housing. That this could happen in one of the world's most advanced cities, in the modern day and age – 1929 – was incomprehensible to Fuller. So he set to work on a variety of inventions that pushed the frontiers of engineering. His plans for levitating① cars and boats that were propelled by opening and closing cones② above them were all part of the learning curve that led to the Manhattan dome project.

The timing for the development of geodesic domes was perfect. Although originally patented in the 1920s by German designer Walther Bauerfeld, Fuller believed he could improve upon them. They were ingenious③, they were strong, but Fuller thought that with some changes they could do more for mankind than simply house exhibits, which was Bauerfeld's original purpose. Examining the mathematical and engineering concepts behind the domes, in 1948 he discovered the mathematical formula for the closest packing of spheres. The theory was this. Geodesic domes④ can cover extensive areas without internal supports, and accordingly can contain more space with less material than other structures. That means they are relatively cheap too. Moreover, the bigger they are, the stronger they become, so they can contain a lot of city. Finally, because they keep the elements out and the heat in, they can be made to adopt

注释

① levitate ['lɛvɪˌteɪt] *v.* 飘向空中
② cone [kəʊn] *n.* 圆锥体
③ ingenious [ɪn'dʒiːnjəs] *adj.* 灵巧的，新颖的
④ geodesic domes *n.* 网格球形穹顶

中心街区，所覆盖的城区气候自成一体，没有白雪飘飘的漫长冬天，雨也只有在需要时才会下。

这些伟大的想法来自一位智力超群的人，他后来成为世界高智商人士俱乐部门萨协会的主席。巴克敏斯特·富勒想让世界变得更美好的愿望背后还有他的个人动机，那是他亲身经历的一场悲剧。他的女儿在4岁时由于住房条件差而患病夭折，富勒实在想不通这样的事竟然会发生在世界上最发达的城市，而且发生在现代的1929年。于是他开始努力工作，发明了许多推动工程学前沿的东西。他曾设想通过开关车、船上面的圆锥体推动它们飘在空中，这是产生曼哈顿穹顶项目的学习曲线的一部分。

富勒开发网格穹顶正赶上了好时机。虽然德国设计师瓦尔特·鲍尔菲尔德（Walther Bauerfeld）在20世纪20年代就已申请了这项专利，但富勒认为自己可以对此作出一些改进。鲍尔菲尔德的穹顶最初是用来容纳展品的，它设计巧妙、结实坚固，但富勒认为如果加以改进，这种建筑能为人类做出更多贡献。1948年，通过研究穹顶的建筑数学和工程学原理，他发现了球体最紧密堆积的数学公式。其理论如下：网格穹顶能在没有内部支撑的条件下覆盖广阔的区域，所以与其他结构相比，这种建筑结构能用较少的材料容纳较多的空间，这就意味着降低了成本。此外，穹顶体积越大，就越坚固，这样就能容纳很大一片城市区域。最后，由于穹顶能阻挡外界的恶劣天气，因此可以保持室内的温度，用于形成内部的小气候，冷热皆由人设定。

富勒自称是“综合预见性设计科学家”。综合指用最

注释

① anticipatory [æn͵tɪsɪ'peɪtərɪ] *adj.* 期待中的
② tetrahedral [͵tetrə'hiːdrəl] *adj.* 四面体的，有四面的
③ capture ['kæptʃə] *v.* 俘虏，捕获
④ terminology [͵tɜːmɪ'nɒlədʒɪ] *n.* 术语

their own internal microclimate. Hot or cold, you can create the climate you want.

Casting himself as a 'comprehensive anticipatory[①] design scientist' – comprehensive meaning he wanted to benefit the maximum number of people using the minimum amount of resources – domes could fulfil a useful environmental function; improving housing so that others would not have to suffer the death of a child. Fuller looked at inhospitable parts of the globe such as Africa and Antartica and could see a use for the domes. One idea was for a tetrahedral[②] city that would float in Tokyo Bay and house a million people. But it was the Manhattan Dome that captured[③] America's imagination.

But could it be done? An early classroom-based attempt to construct a test dome out of Venetian blinds failed when the dome fell in on itself almost as soon as it was completed, causing cynics to name it the 'flopahedron'. In response, Fuller argued that the collapse was intentional – he wanted to understand the critical point to which he could go whilst in a safe environment. Then came a success: the Ford Motor Company commissioned the first commercial geodesic dome from Fuller in 1953, spanning 93ft at a weight of just 8.5 tons. With a successful project under his belt, demand for geodesic domes suddenly rocketed.

Plans for Manhattan's dome progressed. True, at $200 million it would be costly, but the 4D concept of lifetime's use came into play: over time, living inside a dome that was constructed out of fewer resources than a traditional building would save cash. The climate could also be efficiently managed. Indeed in today's terminology[④], geodesic domes were green, not least because the dome would pay for itself within ten years in the cash saved from clearing away New York's snow alone. Eventually the site was located, spanning '50 blocks of the

少的资源使最多的人受益，而他设计的穹顶具有实用的环境功能，改善了居住条件，从而使其他人不再遭遇孩子病死的噩运。富勒研究了地球上一些环境恶劣的地区，譬如非洲和南美洲，认为这些地区可以使用穹顶建筑。他曾经设想在东京湾上建一座可容纳100万人口漂浮的四面体城市，但真正抓住美国人想象力的还是曼哈顿穹顶。

但这个想法能实现吗？富勒在早期的课堂上尝试用软百叶帘制作了一个实验穹顶，但失败了，穹顶在快要完工时倒塌了，一些爱讽刺别人的人把它称为“仰卧穹顶”。富勒辩解说，他是故意让穹顶塌倒的，这样就可以掌握穹顶处于安全环境时的临界点。成功随之而来：1953年，福特汽车公司向富勒订购了第一座商业网格穹顶建筑。这座穹顶建筑跨距为28.4米（93英尺），重量仅为8.5吨。具备了一个成功项目的经验之后，富勒的网格穹顶订单数量开始飙升。

▲ 巴克敏斯特 · 富勒为纽约设计的穹顶的外观。这个高1600米、宽3000米的穹顶计划采用直升机安装到位。

How Buckminster Fuller's geodesic dome for New York might have looked. The segments for the 1.6km high, 3km wide dome were to be flown into place by helicopters.

曼哈顿穹顶计划有了进展。2亿美元的造价确实很贵，不过一旦建好，终身使用的四维概念开始发挥作用：随着时间推移，人们会发现，住在这样一个使用资源比传统建筑少的穹顶里，会节省许多成本。此外，穹顶内部的气候也能得到有效控制。按现在的话来说，网格穹顶属于绿色建筑，至少穹顶建好后，10年内清扫纽约市积雪所省下的钱就可以抵消最初的高

upper Manhattan skyscraper city' from the East River to the Hudson at 42nd Street, and north and south from 62nd Street to 22nd Street. Sixteen large Sikorsky helicopters would fly all the segments① into position for the 1.6km high, 3km wide dome. It would be constructed within three months.

But construction never took place. The political will was lacking and so was broad residential② and commercial support. It's easy to get excited by a domed city when you live miles away, but less so if someone wants to encase③ your home or business in one. For many people, the whole concept was too futuristic. And although he went on to be nominated for the 1969 Nobel Pace Prize, Fuller's inventions sometimes lacked credibility. His Dymaxion Vehicles, three-wheeled cars that could do a 180° turn in a single swift maneouvre④, which was handy for getting in a parking space but little else, caused such traffic chaos that residents pleaded with him to keep off the roads at peak times. Odder still, Fuller claimed that dolphins evolved from humans, which wasn't entirely the prevailing orthodoxy⑤, and that, as captain of 'Spaceship Earth' he would care for everyone on board, combining the role with that of something he called Guinea Pig B, just one of the nicknames he liked to be known by. His Synergetic Geometry system, based on 60° angles rather than the traditional 90°, failed to catch on too. But, most crucially, in addition to the $200 million bill and the resistance of Manhattan residents to seeing their homes encased, geodesic domes had flaws.

For a start, they weren't entirely waterproof, which while fine for encasing cities, didn't do much for the argument that they could control the climate. Next it proved difficult to subdivide the interior⑥. Much of the available space was at a high level, which was impractical or expensive to use fully for other purposes. Even Fuller's own dome, one he had built as

注释

① segment ['sɛgmənt] *n.* 部分
② residential [ˌrɛzɪ'dɛnʃəl] *adj.* 住宅的
③ encase [ɪn'keɪs] *v.* 包，围
④ manoeuver [mə'nuːvə] *n.* 作战行动，军事演习，策略
⑤ orthodoxy ['ɔːθəˌdɒksɪ] *n.* 正统观念
⑥ interior [ɪn'tɪərɪə] *n.* 内部，里面

额建筑成本。最后穹顶的地点也选好了，它将覆盖从东河到第42街哈德逊河“上曼哈顿摩天大楼市区50个街区”，南北跨越第62街和第22街。16架西科斯基直升机将把这座高1600米、宽3000米的穹顶的各个部分搬运到安装地点。整个工程在3个月内即可完成。

然而工程始终没有开始。政府官员并不支持这项计划，大部分当地居民和商人也不赞同。试想你住在离一个有穹顶建筑的城市数英里以外的地方，对于这样一个工程你一定会很激动，可是如果有人想把你家或者你的公司放在这样一个穹顶建筑里，你也许就不会那么激动了。对于许多人来说，整个构想有点过于未来主义了。虽然富勒后来获得1969年诺贝尔和平奖提名，但他的发明有时却缺乏可信度。他发明了Dymaxion车（Dymaxion为最大限度利用能源的意思），这是一种可以通过一键操作即可转弯180度的三轮汽车，除了方便停车入库外，没有别的优点；这种车还会造成交通混乱，当地居民甚至恳求他在交通高峰时不要开这种车上路。更奇怪是，富勒声称海豚是从人类进化来的，这完全与正统学说背道而驰；还有作为“地球太空船”的船长和“豚鼠B”（他喜欢人们叫他的绰号），他会照顾船上的每一个人。他创建的基于60度角而非传统90度角的融合几何系统，也未能得到人们的广泛接受。然而除了2亿美元的工程造价和曼哈顿居民的抵制之外，曼哈顿穹顶计划未能实施的关键原因是网格穹顶存在缺陷。

首先，这种穹顶并不完全防水，虽然这点于所容纳的城区并无大碍，但不利于控制内部小气候。其次，要将穹

a home and where he lived for some time, leaked. And when builders set to work on repairs, they ended up burning down the whole building.

In the end, the project never progressed beyond Fuller's imagination and the drawing board. It did result in acres of positive press coverage and increased demand for domes around the world. The Pentagon invited Fuller to design protective housing for radar equipment while Soviet President Khrushchev ordered a dome for a Moscow fair and wanted Fuller to teach engineering to Soviet engineers. And although the Manhattan domed city didn't come to pass – or at least it hasn't yet – the prevalence of geodesic domes today is just one of Richard Buckminster Fuller's considerable scientific, environmental and social legacies.

顶建筑内部空间进行分割并不容易。许多空间位置很高，很难充分利用或低成本使用。富勒曾给自己设计了一个穹顶建筑的家，还在里面住过一段时间。但这个穹顶建筑也出现了漏水现象。建筑工人对这个房屋进行修葺未果，最后放火烧了整个建筑。

最终，曼哈顿穹顶项目没能进行下去，只是永远留在了富勒的头脑里和绘图板上。不过，它还是带来了一些正面的媒体报道，并增加了全世界对穹顶建筑的需求。五角大楼邀请富勒设计雷达设备的保护罩壳；苏联首脑赫鲁晓夫为莫斯科博览会专门订购了一个穹顶，并希望富勒为苏联工程师教授工程学课程。虽说曼哈顿穹顶城没有成功——或者至少说尚未成功，但今天我们常见的网格穹顶建筑却是富勒为科学、环境和社会留下的重要遗产。

BANNED

THE BRAND NEW CONTINENT OF ATLANTROPA

The idea that one could create an entirely new continent surely verges[①] on the megalomaniacal[②], especially being conceived, as it was, by a German in an era which saw the rise of fascism[③]. But the architect Herman Sorgel concluded that his colossal[④] geopolitical scheme could solve the world's problems – or at least Europe's. He also believed that the technology existed to make his vision happen, even if it would take more than a century to complete.

Sorgel wasn't the only one. The German architect received many plaudits[⑤] from other architects for his plan which involved building a massive dam[⑥] across the Strait of Gibraltar, causing parts of the Mediterranean[⑦] to dry up, creating a wholly new coastline and opening up vast new areas to inhabitation, cultivation and industry.

注释

① verge [vɜːdʒ] *v*. 接近，濒临
② megalomaniacal [ˌmegələʊmə'naɪəkəl] *adj*. 夸大狂的
③ fascism ['fæʃɪzəm] *n*. 法西斯主义
④ colossal [kə'lɒsəl] *adj*. 巨大的
⑤ plaudits ['plɔːdɪtz] *n*. 喝彩，赞扬
⑥ dam [dæm] *n*. 水坝
⑦ mediterranean [ˌmɛdɪtə'reɪnɪən] *n*. 地中海

项目遭禁

亚特兰特洛帕新大陆

一个人能够创造一个全新的大陆，这个想法可真有点儿狂妄自大，特别是当它来自一个法西斯时代的德国人时更是如此。但建筑师赫尔曼·索格尔（Herman Sorgel）认为，他那宏伟的地缘政治计划能够解决世界问题，或者至少解决欧洲问题。他还相信现有的技术能实现他的想法，不过可能要用一个世纪的时间来完成。

索格尔不是唯一抱有这种想法的人。这位德国建筑师的计划得到了其他建筑师的热情赞扬。他的计划包括在直布罗陀海峡上修筑一个巨大的水坝，使地中海部分区域干涸，从而创造出一个全新的海岸线和广阔的新陆地供人居住、耕作和生产。索格尔雄心勃勃，想把欧洲和非洲合并成一个卓越的新大陆，他将其命名为亚特兰特洛帕。而《星际迷航》的作者吉恩·罗登伯里（Gene Rodenberry）似乎也受到了类似想法的启发。在他1979年出版的书中，他让科克船长（Captain Kirk）站在地中海上一座用于发电的大坝之上，这个大坝与索格尔的想法不谋而合。

注释

① fuse [fju:z] *n.* 保险丝
② sketch [skɛtʃ] *n.* 草图，略图，素描
③ refined [rɪ'faɪnd] *adj.* 去掉杂质的，精炼的
④ fulcrum ['fʊlkrəm] *n.* 支柱

Sorgel's ambition was to fuse[①] Europe and Africa into a brave new land, which he named Atlantropa. Even Gene Rodenberry, the creator of the TV series *Star Trek*, was seemingly inspired by a similar notion. In his 1979 book he has Captain Kirk standing on a huge structure damming the Mediterranean to produce hydroelectric power, just like Sorgel.

Thankfully, the Nazis, who rose to power in the 1930s as Sorgel was busy sketching[②] out his master plan, didn't seize on the idea. They were too busy eyeing up the territory that already existed to take Herman, a pacifist, too seriously. His works were banned by Hitler's government in 1942. Sorgel, who produced a staggering 1,000 publications on Atlantropa during his lifetime, thought his concept would help bring peace to Europe, which had already been ravaged by the First World War, as well as provide employment and ease the pressure from its expanding population. But many have branded Sorgel's Eurocentric views plain racist – especially his plans for Africa. In his world view, the supremacy of Europe was what mattered. He feared a future where the globe would be divided up into three huge competing blocks: America, Europe and Asia, with Europe potentially weaker than the other two.

So just what did Sorgel propose? Working as a sometime architect, writer and teacher he first started work on the Atlantropa idea in 1928 and refined[③] his ideas over the years. The fulcrum[④] of his plan, it emerged, was a massive dam stretching from Gibraltar across to Morocco, closing the narrow gap which separates Africa and Europe and links the Atlantic Ocean to the Mediterranean. The structure, which would be 18 miles long, and would take a million workers ten years to build, would be crowned with a 1,300ft-high decorative tower designed by fellow German architect Peter Behrens. Another dam would be placed at the other end of the

值得庆幸的是，20世纪30年代在索格尔忙着制定总体规划时，刚上台的纳粹并未注意到他的想法。他们忙着对现有的领土虎视眈眈，并没把像赫尔曼这样的一个和平主义者放在心上。1942年，索格尔的作品遭到希特勒政府的查禁。他一生中关于亚特兰特洛帕的著作多达1000部，令人惊讶。他认为他的构想能给饱受第一次世界大战摧残的欧洲带来和平，还能提供就业机会，缓解人口扩张的压力。但是许多人批评索格尔这个以欧洲为中心的构想具有明显的种族主义，特别是他关于非洲的计划。在他的世界观里，欧洲至上才是主要问题。他担心世界将来会分成三大竞争板块：美洲、欧洲和亚洲，而欧洲有可能成为三者中最弱的一方。

Big Dam to Water Sahara

▲ 德国建筑师赫尔曼·索格尔正在设计横跨直布罗陀海峡的大坝图纸，这是他创造亚特兰特洛帕新大陆的梦想的一部分。

German architect Herman Sorgel working on plans for a dam spanning the Strait of Gibraltar; part of his dream to create the new continent of Atlantropa.

那么，索格尔到底在他的计划中写了些什么？1928年，建筑师、作家和教师索格尔开始了他的亚特兰特洛帕计划，之后多年他不断改进自己的构想。他计划的核心支柱，是在横跨直布罗陀海峡至摩洛哥的区域修建一座巨大水坝，以填补分割非洲和欧洲以及连接大西洋和地中海的缺口。这座建筑长约29米（18英里），将耗费100万人力，历时10年完成，大坝顶部将立有由他的德国建筑师同事彼得·贝伦斯（Peter Behrens）设计的高约396米（1300英尺）的装饰塔。另一座水坝将建在地中海的另一端，那里是与黑海相连的狭窄的达达尼尔海峡。大坝边上将修建运河，

Mediterranean, where it is connected to the Black Sea at the narrow Dardanelles. Canals would be built beside the dams so that shipping could still pass through.

Sorgel said that these dams would reduce a natural inflow of water into the Mediterranean from the Atlantic and Black Sea. Not only would the dams produce huge amounts of hydroelectric power once this supply of water was cut off, the continuing evaporation① in the Mediterranean would have the effect of gradually lowering the sea level. Within 100 years, said Sorgel, it would be 330ft lower. Eventually this would leave 220,000 square miles of virgin land ripe for exploitation and development. Sicily②, for example would now be joined by a land bridge to mainland Italy, Corsica would be joined to Sardinia, while the Adriatic Sea③ would end up almost completely drained. Interestingly, the city of Venice would get its own special dam, effectively turning its famous lagoon into a lake.

Large parts of what Sorgel saw as an unproductive Africa would be flooded with another dam across the Congo River. It would involve moving two million inhabitants but Sorgel thought it was worth it. Meanwhile, the dunes④ of the Sahara could be reclaimed for farming. In July 1933 the American *Popular Science* magazine reported:

> Turning the Sahara Desert into blossoming farm land, with water drained⑤ from the Mediterranean sea, is the ambitious project for which Herman Sorgel, a German engineer, seeks international support. He proposes to dam the Strait of Gibraltar and then cut a canal to flood portions of the Sahara below sea level. Evaporation from the inland lake thus formed would produce rain clouds and water a vast area, he maintains.

注释

① evaporation [ɪˌvæpəˈreɪʃən] *n.* 蒸发

② Sicily [ˈsisili] *n.* 西西里岛（意大利一岛名）

③ Adriatic Sea 亚得里亚海（地中海的一部分，位于意大利东海岸和巴尔干半岛之间）

④ dune [djuːn] *n.* 沙丘

⑤ drain [dreɪn] *v.* 使流走，流走

方便船只通行。

索格尔称，这些水坝将减少从大西洋和黑海自然流入地中海的水量。大坝不仅会带来巨量的水电，而且地中海的海水供给一旦被切断，海水的持续蒸发将会使海平面逐渐下降。索格尔认为，在100年内，地中海的海平面将降低约101米（330英尺），这将产生约57万千米^2（22万英里^2）的处女地供开发和利用。譬如，到那时，一个连接西西里岛和意大利的大陆桥将出现，科西嘉岛将与撒丁岛相连，而亚得里亚海将完全干涸。有趣的是，威尼斯城将有自己的水坝，那里有名的潟湖将会变成湖泊。

而被索格尔视为非洲大块的不毛之地将被刚果河上的一个水坝淹没。这意味着200万居民将不得不搬走，但索格尔认为这样做是值得的。同时，撒哈拉沙漠的沙丘将被重新开垦用于耕作。1933年7月，美国杂志《大众科学》报道说：

> 在地中海的滋润下，撒哈拉沙漠将成为生机勃勃的农田——这是德国工程师赫尔曼·索格尔一个雄心勃勃的项目，他正在寻求国际上的支持。他提议在**直布罗陀海峡**上建一个大坝，然后修一条运河，将撒哈拉沙漠低于海平面的地区淹没。他说，来自内陆湖的水蒸气将产生雨云，浇灌一片广阔的地区。

注释

直布罗陀海峡是沟通地中海与大西洋的海峡，位于西班牙最南部和非洲西北部之间，长58千米。最窄处在西班牙的马罗基角和摩洛哥的西雷斯角之间，宽仅13千米。

在之后的几年里，索格尔更详尽地阐述了他的理论。根据他的计划，在突尼斯与西西里岛之间将修建一座桥

Over the years Sorgel elaborated further. A bridge would be built between Tunisia and Sicily, easing travel between the two continents. The new dams, together with new power plants dotted around the Mediterranean, would produce a useful continent-wide power grid too. Sorgel even sought to collaborate with the German-Jewish architect Erich Mendelsohn over a new coastline for a potential future Jewish state in Palestine, years before Israel finally got statehood[①] in 1948.

There were some big, practical problems with his ideas, not least that some of the Mediterranean's biggest ports would now be stranded inland. But in the 1930s many of his peers thought Sorgel was on to something. After all, that mammoth[②] undertaking in America, the Hoover Dam, was completed in 1936. In addition, thousands of square miles of fertile land had already been reclaimed in the Netherlands when the Zuidersee, a shallow bay in the centre of the country, was dammed. Neither did many, back then, easily baulk[③] at the idea of displacing millions of people, in what they saw as the interests of social and technological progress.

His plan does seem hopelessly grand. But Sorgel's fear that Europe would decline while America and Asia became more powerful has, to some extent, been borne out in the decades which have followed. Europe has certainly become more interdependent as he knew it would have to be. His fear that fossil fuels would begin to run out and that humankind would need renewable energy sources also seems forward thinking. As we know, Sorgel's ideas came to nothing, though interestingly the idea of building a bridge or tunnel between Spain and Morocco has recently been discussed between the two countries.

Sorgel, of course, was a fantasist[④]. And despite spending hours coming up with great architectural schemes the only structures he ever built, according to reports, were a few houses.

注释

① statehood ['steɪthʊd] *n.* 独立国家地位

② mammoth ['mæməθ] *adj.* 巨大的，艰巨的

③ baulk [bɔːk] *v.* 阻止，反对

④ fantasist ['fæntəsɪst] *n.* 幻想者

梁，方便两个大陆之间的交通。而新建的水坝将与遍布地中海的新发电厂一起，形成一个电网，为整个大陆提供电力。索格尔甚至寻求与德裔犹太人建筑师埃里希·门德尔松（Erich Mendelsohn）合作，为将来在巴勒斯坦建立一个犹太国家规划一条新的海岸线。这比以色列最终在1948年建国早了好几年。

他的想法有几个大的实践性困难，其中之一是地中海的一些最大的港口会被遗留在内陆。但是在20世纪30年代，索格尔的许多同行都认为他的计划有一定可行性。毕竟，美国建造的那座庞然大物——胡佛大坝——在1936年竣工了。再说，当须德海这个位于荷兰中心的浅海湾的出口被筑起大坝之后，荷兰便通过围海造地索回了数千平方英里的肥沃土地。那时候，多数人并不会因为想到要把数以百万计的人口驱离家园就犹豫不决，相反，他们认为这是为了社会和技术的进步。

的确，索格尔的计划看起来太大了。不过，索格尔认为欧洲将衰落，美洲和亚洲将更加强大，这个担心在几十年后或多或少得到了证实。欧洲国家之间变得更加相互依赖，索格尔也预见到了。他担心化石燃料将耗尽，人类将需要可再生能源，也很有前瞻性。正如我们所知，索格尔的计划最终没能实现，不过有意思的是，西班牙和摩洛哥最近刚讨论过在两国之间修建一座桥梁或一条隧道。

当然，索格尔是位幻想家。据一些报道称，他除了花大量时间来构思宏伟的建筑蓝图，真正完成的建筑仅是几座房子。甚至在第二次世界大战以后，他的想法还备受关注，在德国有一个亚特兰特洛帕研究所，国外那些感兴

注释

① dignitary ['dɪgnɪtərɪ] *n.*（政府或教会的）显要人物

Even after the Second World War, his idea received plenty of attention and there was an Atlantropa Institute in Germany visited by foreign dignitaries[①] keen to know more. For a man with such an exalted scheme, poor old Herman had a humble end. While cycling to give a lecture on the project he was run over by a car, and died on Christmas Day 1952. The idea of Atlantropa seems to have died with him.

趣的达官贵人曾去访问。赫尔曼的理想如此崇高，他的死却极其卑微。在骑自行车去做一个关于这个项目的讲座途中，他被一辆汽车碾过。他于1952年的圣诞节那天去世，亚特兰特洛帕计划也随他一起消失。

ABANDONED

A NATION BUILT ON SAND

① Ealing ['i:liŋ] *n.* 伊林（英格兰东部城市名）
② disgruntled [dɪs'grʌntəld] *adj.* 生气的，不满的
③ mogul ['məʊgəl] *n.*（尤指新闻、影视界的）大人物，大亨

In the 1949 Ealing[①] comedy film *Passport to Pimlico*, a band of Londoners revel in the discovery that their part of the city isn't British at all but legally part of France. The movie's story taps into a deep desire among many to be free of the state they find themselves in. History has, of course, seen the rise of many virgin nations from the desire of peoples to be independent. Since 1990 the number of new states has grown by at least thirty. But, in a crowded world, it's been getting pretty difficult to start your own sovereign state from scratch. That hasn't stopped some people trying. A name for the more eccentric bids to state-building has even been coined: micronations.

While onlookers might find many of these attempts ridiculous, their proponents, often disgruntled[②] groups or individuals, are often deadly serious. None more so than those who led the bid to found the Republic of Minerva; a nation constructed where there had once only been water. The main motivator behind it was Lithuanian-born Michael Oliver, a Jewish property mogul[③] from Nevada. But this was not merely some late-night, bar-room fantasy. Oliver's dream of a

建在沙上的国家

在1949年的伊林喜剧电影《买路钱》中，一群伦敦人发现他们所在的城区并不属于英国，而是法国的合法领地，他们为此而狂欢。这部影片挖掘了许多人的一个深层愿望，就是摆脱他们所在的国家。当然，历史上也确实因为一些民族要求独立而诞生了许多新国家。自1990年起，至少产生了30个新生国家。但是，要在拥挤的地球上从零开始建立起属于自己的主权国家，可不是一件容易的事儿。不过，这并没有阻止一些人去尝试。人们甚至还给这种古怪的建国现象发明了一个新词——微型国家。

旁人也许会觉得这种尝试荒唐可笑，但倡导者们（常常是一些对社会不满的群体或个人）却往往把它当回事。一个典型的例子就是那些想建立密涅瓦共和国的人，这个国家在创建的时候只有海水。迈克尔·奥利弗（Michael Oliver）是这个国家的主要发起者，他出生于立陶宛，是一位来自内华达州的犹太房地产大亨。然而，这个愿望不只是产生于深夜酒吧间的一个幻想。奥利弗无税的自由意

libertarian state, with no taxes, did for a short time become a reality.

The nation's nickname, 'Land of the rising atoll[1],' gave a clue as to just how this country was to come about. It would be made from sand and gravel[2] and be built on existing, unclaimed reefs[3] that lay just a metre below the surface of the ocean. Located some 270 miles south west of Tonga, the Minerva reefs had been named after a whaling ship of the same name that had been wrecked[4] there in 1829. The Republic of Minerva was also destined to flounder but not before a band of desperados[5] had made serious practical attempts to set up an island nation that was ultimately designed to home 30,000 people, boast its own currency, constitution and capital.

In 1971, Oliver, backed by 2,000 Americans hoping to establish a rule-free capitalist utopia, arranged for barges[6] loaded with sand to arrive from Australia. It was duly dumped[7] and gradually the level of the island was built up, enough for a stone platform to be constructed. On 19 January 1972, a flag bearing a golden torch on a bright blue background was raised on this lonely spot in the middle of the Pacific Ocean. Coins were produced bearing the bust of the Roman goddess Minerva, along with the latitude[8] and longitude[9] of the proposed nation. Bizarrely the 10,500 Minerva dollar coins that were minted[10] came with a value of 35 Minervan dollars. This did, at least, make them something of a valuable curiosity for collectors.

In February 1972, Morris C. Davis was elected as provisional president of the Republic of Minerva. Declaring the aims of the new country he said: 'People will be free to do as they damn well please. Nothing will be illegal so long it does not infringe on the rights of others.'

This certainly wasn't the sort of light-hearted episode that featured in Passport to Pimlico. It had got the neighbouring

注释

① atoll ['ætɒl] *n.* 环形珊瑚岛
② gravel ['grævəl] *n.* 砂砾
③ reef [ri:f] *n.* 礁
④ wreck [rɛk] *v.* 摧毁
⑤ desperado [ˌdɛspə'rɑ:dəʊ] *n.* 亡命徒
⑥ barge [bɑ:dʒ] *n.* 平底载货船
⑦ dump [dʌmp] *v.* 扔下，倾倒
⑧ latitude ['lætɪˌtju:d] *n.* 纬度
⑨ longitude ['lɒndʒɪˌtju:d] *n.* 经度
⑩ mint [mɪnt] *v.* 铸造

志主义国家梦想的确成了现实，尽管只有很短一段时间。

这个国家又被称为“升起的环礁国”，从名字上我们就可以猜测出它是如何产生的。它由沙子和碎石筑成，建在海平面1米之下、无人认领的礁石之上。这些礁石就是位于汤加西南部约435米（270英里）的密涅瓦礁脉，是以1829年在那里失事的一艘捕鲸船命名的。密涅瓦共和国之所以遭受厄运，还因为那伙亡命之徒动真格地要建一个3万人的岛国，还要创造自己的货币、宪法和首都。

1971年，在2000名希望建立一个没有约束的资本主义乌托邦的美国人的支持下，奥利弗安排好从澳大利亚出发的载沙驳船。沙子被如期倒入海中，渐渐地出现了一个小岛，大小足够修建一个石头平台。1972年1月19日，一面旗帜在太平洋中间这个孤零零的小岛上升起，旗帜明亮的蓝色背景上有一个金色火炬。他们还铸造了货币，上面印有罗马女神密涅瓦的半身像以及该国所处的经度和纬度。奇怪的是，所造的10500枚密涅瓦1元硬币，面值却为35密涅瓦元。至少，这确实使它们成为收藏者有价值的珍品。

1972年2月，莫里斯·C. 戴维斯（Morris C. Davis）被选为密涅瓦共和国临时总统。在宣布这个新国家的宗旨时，他说：“人民绝对可以做他们想做的事情。只要不侵犯他人的权利，没有什么事情是违法的。”

这可不是电影《买路钱》中的喜剧片段。它让邻国汤加人感到很不舒服。1972年，汤加国王决定采取行动。根据1972年6月15日的《汤加政府公报》，国王陶法阿豪·图普四世发布公告，密涅瓦礁脉“12英里半径范围内的岛屿、岩石、礁脉、涨滩及水域”均属于汤加王国。

Tongans feeling uncomfortable and, in June 1972, the Tongan king decided to act. On 15 June, the Tongan Government Gazette reported that His Majesty King Tāufa'āhau Tupou IV had issued a proclamation[1] that 'the islands, rocks, reefs, foreshores and waters lying within a radius[2] of twelve miles' of the reefs were part of the Kingdom of Tonga. On 21 June, the king set out on board the royal yacht[3] Olovaba for the reefs. With him were members of the Tonga Defence Force, a convict work detail and a four-piece brass band in order to enforce the claim. The republic's flag was lowered by the king who clambered[4] onto the island himself. Quite an achievement given the fact that his highness was also his greatness; he weighed in at 350lbs.

But what about the other neighbouring nations? After all, the territory was actually a long way from Tonga. If there was land here for the taking wouldn't other countries fancy a piece of it too? As it turned out, the Tongans needn't have feared that the incident was about to spark a full-scale pan-Pacific conflict. A few months later The South Pacific Forum, made up of heads of government from the Pacific island states, recognised the Minerva Reefs as Tongan territory. For the Republic of Minerva, it was pretty much the end of the line as the nation slowly sank beneath the waves once more.

Oliver's attempts to start another nation certainly didn't end there, however. His Phoenix Foundation backed attempts in 1973 to declare a separate country in the island of Abaco, part of the Bahamas, and even trained a militia to take it by force. In 1980 the same organisation backed The New Hebrides Autonomy Movement, aiming to break free from the new nation of Vanuatu in the Pacific on the island of Espiritu Santo. It was quashed[5] when the Vanuatu government organised for a battalion[6] to occupy it.

注释

① proclamation [ˌprɒkləˈmeɪʃən] *n.* 声明
② radius [ˈreɪdɪəs] *n.* 半径
③ yacht [jɒt] *n.* 赛艇，游艇
④ clamber [ˈklæmbə] *v.*（手脚并用，费劲地）爬
⑤ quash [kwɒʃ] *v.* 撤销，废止
⑥ battalion [bəˈtæljən] *n.* 营（由三个或以上的连组成的）

6月21日，国王乘皇家游艇“奥罗法巴号”前往密涅瓦礁脉，随行助威的还有汤加国防军部队、一个犯人劳动小组和一支四人铜管乐队。国王亲自艰难地爬上小岛，将密涅瓦共和国的旗帜降下。国王大驾亲临，这可是件了不得的事，而整个过程由体重约159千克（350磅）的国王亲自完成，这更是个壮举。

其他邻国又是怎么想的呢？毕竟，这块土地距离汤加其实很远。对于一块没被认领的土地，难道其他国家就不会认为这也是它们的吗？事实上，汤加人无须担心这件事将激起一场全面的泛太平洋冲突。几个月后，由太平洋岛国政府首脑组成的“南太平洋论坛”承认密涅瓦礁脉为汤加领土。而对于密涅瓦共和国来说，当这个国家再次慢慢沉入海浪之中时，故事也就差不多结束了。

不过，奥利弗并未就此停止他的建国尝试。1973年，在他的火鸟基金会的支持下，奥利弗在阿巴科岛（巴哈马群岛的一部分）宣布建国。他甚至还训练了一支民兵队伍，准备以武力拿下这座岛。1980年，这个基金会还支持了新赫布里底斯群岛自治运动，试图将太平洋圣埃斯皮里图岛从新国家瓦努阿图独立出来。结果瓦努阿图派了一个营占领了该岛，自治运动遭到镇压。

除了奥利弗梦想创造一个逃税人天堂，不少人都尝试了创建微型国家。雷东达王国是最早成立的微型国家之一。1865年，一名叫马修·道迪·谢尔（Matthew Dowdy Shiell）的商人登上一座无人居住的加勒比海岛，宣布这是他的王国。150年后，人们仍在争夺这个岛屿的主权。在英国附近也有这样的事情。1967年9月2日，英军前少

Oliver's dream of a tax loathers' paradise was just one of scores of attempts to create micronations. One of the original examples was The Kingdom of Redonda. In 1865 a trader called Matthew Dowdy Shiell landed on the uninhabited Caribbean island and declared it his own kingdom. One hundred and fifty years later the island's ownership is still contested. There have been attempts closer to Britain, too. Sealand, founded by former British army major Paddy Roy Bates on 2 September 1967, was based on a derelict[①] anti-aircraft tower 6 miles off the coast of Suffolk in the North Sea. In the past, British courts have ruled that they didn't have jurisdiction[②] over the platform. His principality went as far as issuing passports, hurriedly revoked[③] when crooks[④] forged them – one emerged during the investigation of fashion designer Gianni Versace's death in 1997.

There aren't many free locations for anyone wanting to follow in the footsteps of these wannabe[⑤] nation builders. Most of the world's landmasses are very much spoken for. But there is a part of the world that remains technically unclaimed – Marie Byrd Land in Antarctica, covering a massive 1,610,000 square kilometres. Until now the extreme climatic conditions mean no country has attempted to lay claim to this area as they have to other parts of the icy continent. But with global warming in full swing, now may be just the time to plant your flag.

注释

① derelict ['derɪlɪkt] *adj.* 废弃的
② jurisdiction [ˌdʒʊərɪs'dɪkʃən] *n.* 司法权，管辖权
③ revoke [rɪ'vəʊk] *v.* 撤销，废除
④ crook [krʊk] *n.* 无赖，恶棍
⑤ wannabe ['wɒnəˌbiː] *n.* 效颦者，极力模仿他人者

校派迪·罗伊·贝茨（Paddy Roy Bates）成立了西兰公国。它建在北海一个废弃的防空塔上，距萨福克海岸约10米（6英里）远。过去，英国法庭判决英国对这个防空塔没有管辖权。贝茨的公国后来还签发了护照，可当发现骗子伪造了该国的护照时——假护照是在调查时装设计师杰尼·范思哲（Gianni Versace）1997年死因时被发现的，就急匆匆地把护照撤销了。

现在，地球上已经没有多少自由的土地供人们去创建微型国家了。全球大部分大块陆地已物有所属。但是，从理论上说，地球上还有一块陆地无人占有——南极洲的玛丽·伯德地，它的面积有161万平方千米之广。虽然这块冰雪大陆的其他地方已被不同国家宣布拥有主权，但到目前，还没有哪个国家认领这块气候极其恶劣的土地。不过随着全球气候变暖，现在也许到了在那里插上你的旗帜的时候了。

REJECTED

PAXTON'S ORBITAL SHOPPING MALL

注释

① prefabricated [priːˈfæbrɪkeɪtɪd] *adj.*（建筑物）预制的，组装的

② snap [snæp] *v.* 使咔嚓折断，咔嚓折断

③ moot [muːt] *v.* 提出（计划、想法或主题）供讨论

No one could call Joseph Paxton a failure. He was the man who designed the Crystal Palace – that glorious glass hall which hosted the Great Exhibition in 1851. Paxton's magnificent structure, erected in London's Hyde Park, was visited by six million people during the event. It was certainly much more elegant than most of the other designs which had been on the table. They included an enormous brick building with an iron dome, suggested by none other than Isambard Kingdom Brunel. In fact all of the 245 designs which had been received for the proposed exhibition hall following an international competition were deemed so impractical that when Paxton published his own sketches for a prefabricated① glasshouse in the *Illustrated London News* at the last minute, his scheme was snapped② up.

The hall, later dubbed the Crystal Palace by the satirical magazine *Punch*, was nearly 2,000ft long, 408ft wide, 108ft high and made out of 4,500 tons of iron along with nearly 300,000 sheets of glass. It was put up in just eight months from August 1850. Amazingly, it was only ever meant to be temporary. When the exhibition came to an end no one quite knew what to do with it. Paxton himself mooted③ that it stay where it was

项目遭拒

帕克斯顿的轨道购物商场

没人会说约瑟夫·帕克斯顿（Joseph Paxton）是个失败者。他设计了水晶宫——那幢举办过1851年万国工业博览会的华丽的玻璃屋。它矗立在伦敦海德公园，博览会期间共有600万人出入这座宏伟巨构。在所有提交讨论的设计方案中，帕克斯顿的设计无疑是最典雅的。在这些方案中，有著名设计师伊桑巴德·金德姆·布鲁内尔（Isambard Kingdom Brunel）的铁穹顶砖楼。事实上，这场国际比赛征集的所有245个设计方案都被认为不实用，因此在评选的最后一刻，当帕克斯顿在《伦敦新闻画报》上发表了他的预制玻璃房草图时，他的方案立即被采纳了。

这栋建筑长约610米（2000英尺），宽约124米（408英尺），高约33米（108英尺），共耗费4500吨钢材和近30万块玻璃，后来讽刺杂志《笨拙》给它取了个“水晶宫”的绰号。大楼从1850年8月开始动工，仅8个月就建好了。奇怪的是，它只是一个临时建筑。博览会结束以

注释

① pane [peɪn] *n.*（门、窗上的）一块（玻璃）
② girdle ['gɜːdəl] *n.* 紧身褡
③ encircle [ɪn'sɜːkəl] *v.* 环绕
④ termini ['tɜːmɪnaɪ]（terminus 的复数）terminus ['tɜːmɪnəs] *n.* 终点站
⑤ glaze [gleɪz] *n.* 釉
⑥ propel [prə'pɛl] *v.* 推进
⑦ faltering ['fɔːltərɪŋ] *adj.*（尝试、努力、行动等）犹豫的，蹒跚的
⑧ clog [klɒg] *v.* 堵塞

and be turned into a steamy winter garden. One architect even suggested rearranging the panes[1] of glass to turn it into a huge 1,000ft glass tower.

Eventually, the palace was moved, pane by pane, to Sydenham Hill in South London where it reopened in 1854 and stood overlooking the metropolis until it tragically went up in flames in 1936. Not content at wowing the world with his Crystal Palace, Paxton had another, even grander idea up his sleeve. By the mid-1850s Paxton, who had started out designing gardens, was riding high on the success of the Great Exhibition. He'd been knighted for his efforts and had become an MP representing Coventry. In 1855 he felt bold enough to present to his parliamentary peers a plan for the Great Victorian Way, a project which would, he said, 'make London the grandest city in the world' – and make the Crystal Palace seem trifling. The Great Victorian Way was, in effect, a vast, enclosed, shopping mall combined with a ring road and circular railway system. This 'Grand Girdle[2] Railway and Boulevard Under Glass' would encircle[3] London, above ground, in a 10-mile loop crossing the Thames and back again to join up each of the main railway termini[4]. It would also have a mile-long branch line using a third bridge to take travellers to the Houses of Parliament and Victoria Street.

There would be an elevated glazed[5] central walkway-cum-roadway in the middle for pedestrians, wagons and coaches. A breathtaking eight railway lines would be carried in closed galleries around the outside, catering for both express and local stopping services. The trains were to be propelled[6] by atmospheric pressure, a new, somewhat faltering[7], technique. Paxton said the engineer Robert Stephenson, son of railway pioneer George, had assured him it would work. These new lines would drastically cut travelling times across the clogged[8]

后，没人知道该怎么处置它。帕克斯顿认为，可以把它直接改造成一个湿润的冬季花园。一名建筑师甚至提议将玻璃板重新组合，建成一座高约305米（1000英尺）的玻璃巨塔。

最终，这座宫殿被一块玻璃一块玻璃地移到了伦敦南部的锡德纳姆山。1854年，这座俯瞰伦敦城的水晶宫重新向世人开放，直到1936年不幸被大火烧毁。然而，帕克斯顿并不满足于让世人惊异于他设计的水晶宫，他暗怀另一个更宏大的计划。到19世纪50年代中期，由于万国工业博览会的成功，以花园设计起家的帕克斯顿的事业如日中天。由于成就卓著，帕克斯顿被封为爵士，并成为考文垂国会议员代表。1855年，他觉得时机已经成熟，向国会同事展示了他的维多利亚大道方案。按他的话讲，这个项目将“使伦敦成为全世界最伟大的城市”，在它面前就连水晶宫也显得微不足道。实际上，维多利亚大道是一个巨大的、封闭的大型购物中心，配套建有与之相连的环形公路和铁路系统。这个“笼罩在玻璃下的宏伟的环形铁路和林荫大道”将在地面上形成一个约16千米（10英里）长的环形，围住伦敦。它将跨越泰晤士河，与各个主要火车站相连。它还会有一条约1.6千米（1英里）长的支线，采用第三座桥将乘客送至国会大厦和维多利亚街。

在这座建筑中间将架起一条镶有玻璃的中央人行车行道，供行人、货车和客车通行。外面是一个封闭的长廊，长廊内有令人惊叹的八轨铁路，用于高速列车通行及本地停车服务。火车将由气压驱动，这是一项不太成

city (which did not yet have an underground railway). On the Great Victorian Way you would be able to travel from Bank to Charing Cross in just five to six minutes reckoned Paxton, a feat that can hardly be achieved even today.

Within the arcaded[①] central avenue of the great girdle there would be space for shops, hotels, restaurants and even houses. Paxton saw it as a kind of modern version of the medieval London Bridge which, in its time, once had houses and shops along it. All 10 miles of his structure would be 72ft wide and 108ft high and, like the Crystal Palace, would be built from mainly iron and glass, with a great glazed roof all the way round. Paxton's love of glass was fuelled both by his early work on large garden conservatories and by the fact that its price had fallen, making it cheaper to use in large building projects.

When he appeared in front of a select committee in the summer of 1855 to be quizzed on the proposal, Paxton was keen to point out that no 'important street' or 'valuable property' would be knocked down and that his great girdle wouldn't obstruct any existing major thoroughfare[②]. He even claimed that the residences that were to be embedded[③] in the structure would 'prevent many infirm persons from being obliged to go into foreign countries in the winter'. In an article that year the *Civil Engineer and Architect Journal* refers to the Great Victorian Way as The Paxton Arcade and celebrates its 'brightness of light and immunity[④] from weather as a promenade[⑤], and drive, in summer and winter'.

The cost was eye-watering though. His Crystal Palace had cost only £79,800 to build. The Great Victorian Way, admitted Paxton, would need an outlay[⑥] of £34 million. In twenty-first-century terms, with inflation taken into account, you're looking at a project costing billions. At this price Paxton knew it would need not only the blessing of government but its financial

注释

① arcade [ɑːˈkeɪd] *n.*（购物）拱廊
② thoroughfare [ˈθʌrəˌfɛə] *n.* 大道，主要大街，通衢
③ embed [ɪmˈbɛd] *v.* 嵌入
④ immunity [ɪˈmjuːnɪtɪ] *n.* 免除，豁免
⑤ promenade [ˌprɒməˈnɑːd] *n.* 散步场所
⑥ outlay [ˈaʊtleɪ] *n.*（必要的）费用

熟的新技术。帕克斯顿说，铁路先驱乔治的儿子工程师罗伯特·斯蒂芬逊（Robert Stephenson）已向他保证这项技术切实可行。这些新线路将大幅减少在拥挤的伦敦城穿行的时间（那时伦敦还没有地铁）。据帕克斯顿估计，经过维多利亚大道从英格兰银行到达查林十字街只需要五六分钟，这在今天都很难做到。

大环路的拱廊中央大道内，留有建造商铺、旅店、餐馆甚至住房的空间。帕克斯顿把它视为中世纪伦敦桥的现代版本，而中世纪的伦敦桥上是有房屋和商店的。这座约16千米（10英里）长的建筑将有约22米（72英尺）宽，约33米（108英尺）高。像水晶宫一样，主要材料为钢和玻璃，全部为玻璃瓦屋顶。帕克斯顿对玻璃的热爱起源于他早期设计的大温室花园，而他喜欢使用玻璃作为建材的另一个原因是此时玻璃已不像以前那么昂贵了，可以用于大型建筑之中。

1855年夏天，成立了一个特别委员会，对帕克斯顿的项目进行质询。帕克斯顿强调，他的大环路工程不会拆除“重要街道”或“有价值的房屋”，而且不会阻塞任何现有的主要干道。他甚至声称，住在大环路内的房屋里，“老弱者无须去国外过冬”。那年，《土木工程与建筑》杂志将维多利亚大道称为“帕克斯顿拱廊”，还称赞由于“采光好，且不受天气影响，这里无论夏天还是冬天，都是散步和兜风的好去处”。

不过这项工程的造价却高得让人想哭。水晶宫只花了79800英镑就建成了，而建造维多利亚大道将需要3400万英镑的支出。考虑通货膨胀因素，这笔钱在21世纪相

注释

① thorny ['θɔːnɪ] *adj.* 带刺的
② ventilation [ˌventɪ'leɪʃ(ə)n] *n.* 通风设备，空气流通
③ slumber ['slʌmbə] *n.* 睡眠
④ whistle ['wɪsəl] *n.*（某物在空中疾驰而过的）呼啸声，（空气、蒸汽通过狭小孔道时的）尖啸声
⑤ hub [hʌb] *n.* 枢纽机场
⑥ dainty ['deɪntɪ] *adj.* 娇小的，漂亮的

backing too, though he proposed that a private company actually carry out the work.

At first his fellow MPs seemed favourable and certainly gave a lot of consideration to the plan. Something like it was needed. Congestion was already a thorny[①] problem in a city which had swelled in the preceding few decades thanks to industrialisation. As Paxton pointed out, it sometimes took longer to cross central London than to travel from the capital to Brighton. And Paxton's plan had been well thought through. He'd worked on ventilation[②] systems, drawn up plans for the route's bridges and even chosen the kinds of tiles he was going to use for decoration. Not everyone thought the Great Victorian Way was a good idea though. The new *Daily Telegraph* newspaper, born in June 1855, stated in a July article that: 'We must protest against being disturbed in our slumbers[③] by a whistle[④] and a roar overhead.'

However favourable some found the plan, the government were ultimately unwilling to take on such a huge financial burden, even for the tried and tested Paxton. Just a few years later the 'roar overhead' began to be replaced by a roar underneath, as the first of the London Underground lines was built, following the same idea of connecting the big railway hubs[⑤] that Paxton had. The great man lived just long enough to see this much less dainty[⑥] version of his vision become reality when The Metropolitan Line opened in 1863.

当于一个几十亿的工程。帕克斯顿明白，这个造价意味着他的工程不仅需要政府的支持，还要有强有力的经济后盾，尽管他那时说已经有一家私营公司开始着手运作此事了。

起初，他的国会同事貌似赞同他的想法，也对这个项目给予了许多关注。伦敦确实需要一个像这样的东西。对于一个由于几十年工业化发展不断膨胀的城市而言，交通拥挤已成为棘手问题。帕克斯顿指出，有时穿越伦敦市中心所花的时间比从伦敦到**布莱顿**还要长。而且，帕克斯顿还非常周密地考虑了他的规划。他研究了通风系统，对途经的桥梁也制订了计划，他甚至选好了装饰瓷砖。尽管如此，也并不是每个人都赞同维多利亚大道的。1855年6月创刊的《每日电讯报》在7月的一篇文章中写道："我们必须反对，当我们睡觉时，头顶上会有汽笛和轰鸣声骚扰我们。"

注释

布莱顿：英格兰南部海滨城市。距伦敦80多千米。

虽然一些人认为这个方案非常有前景，政府却最终不愿承担这样一笔巨额负担，即便设计者是像帕克斯顿这样久经沙场的建筑师。就在几年后，第一条伦敦地铁线建成，"头上的轰鸣声"被地下的轰鸣声所取代，这与帕克斯顿当初想将伦敦主要铁路枢纽连接起来的想法殊途同归。1863年，在这位伟人逝世前不久，大都会线开通了，他目睹了这一远不及他梦想华美的版本成为现实。

FAILED

THE PERPETUAL MOTION MACHINE

One day as Greek mathematician Archimedes[①] sat in his study in the third century BC, a courtier of King Hiero Ⅱ appeared with a newly cast crown, purportedly[②] made of pure gold. But the king was not so sure. Could Archimedes tell whether a corrupt goldsmith[③] had bastardised[④] his crown with silver?

As with many of the best ideas through the ages, the answer came to Archimedes in the bath. The water level, he noticed, rose when he lowered himself into the tub. Ever the mathematician, Archimedes concluded that: 'a body immersed in fluid exerts a buoyant[⑤] force equal to the weight of the fluid it displaces.' By placing the crown in the bath he was thus able to work out the density of the gold in the crown, by dividing its mass by the volume of water it displaced. History doesn't tell us why he felt compelled to get into the bath with the crown, nor whether he was wearing it at the time. But the answer came to him in a flash. Excitedly running through the streets naked shouting 'Eureka'[⑥], he announced the result: the crown was indeed tainted[⑦] by silver. To this principle of water displacement he gave the humble name Archimedes.

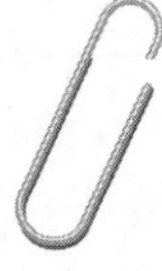

注释

① Archimedes [ˌɑːkɪ'miːdiːz] *n.* 阿基米德，古希腊数学家，物理学家，发明家，学者

② purportedly [pə'pɔːtɪdlɪ] *adv.* 据称

③ goldsmith ['gəʊldˌsmɪθ] *n.* 金匠

④ bastardize ['bɑːstədaɪz] *v.* 腐败，腐坏

⑤ buoyant ['bɔɪənt] *adj.* 快活的

⑥ eureka [jʊ'riːkə] *int.*（欢呼）找到了，有了

⑦ taint [teɪnt] *v.* 玷污

永动机

公元前3世纪的一天，希腊数学家阿基米德（Archimedes）正在书房中坐着，国王希罗二世（Hiero Ⅱ）的一位侍臣拿着一顶新铸的王冠前来找他。据说这顶王冠是用纯金铸成的，但国王并不确信。阿基米德能不能分辨出贪财的金匠有没有在王冠中掺杂了白银呢？

与那个年代许多极妙主意的诞生方式一样，阿基米德的答案也是在浴室里想出来的。他发现当他沉入浴缸时，水位就升高了。最终阿基米德得出结论："人体进入水中产生的浮力等于所排出液体的重量。"于是，他把王冠放入浴缸，通过用王冠的重量除以所排出水的体积，就可以算出王冠的密度。史书没有告诉我们为什么他要戴着王冠进入浴缸，也没有告诉我们当时他有没有戴着王冠，但他就在那一瞬间想到了答案。于是他激动地跳出浴缸，在大街上裸奔，边跑边喊："我发现了！"最终

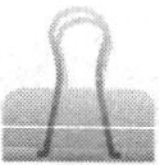

注释

① screw [skruː] *n.* 螺钉
② pump [pʌmp] *n.* 泵
③ bail [beɪl] *n.* 保释金
④ spinning ['spɪnɪŋ] *n.* 纺织
⑤ heretical [hɪ'rɛtɪkəl] *adj.*（信仰）异端的，旁门左道的
⑥ ebb [ɛb] *v.* 退潮
⑦ emulate ['ɛmjʊˌleɪt] *v.* 效仿

For his next trick, Archimedes was to create a screw[1] to help Greece build the largest warship in naval history to date, *The Syracusia*. Because classical Greek shipbuilding wasn't quite up to scratch at the time, the battleship took on water on each voyage, which is never a reassuring prospect for a ship. Archimedes' Screw solved the problem. Through a revolving screw in a cylinder, water could be raised endlessly: the water turns the screw that turns the pump[2], almost but not quite creating its own power to bail[3] out any incoming seawater. Even better, the screw allows water to run uphill. All in all, quite an achievement for a rotating screw.

As the centuries passed, scientists wondered whether the principle could be improved so that machines could work without an external power source. If they could, and if they could do so endlessly, man could put his feet up and let machines run the world. Since then the quest for perpetual motion has driven inventors variously mad, broke or to suicide. And until someone finds a way to break two fundamental laws of physics, perpetual motion will remain impossible. The good news is that that's exactly the kind of challenge that scientists adore.

Renowned inventors, including Richard Arkwright, who came up with the spinning[4] jenny, and father of the railways George Stephenson, worked on the margins of the perpetual motion concept. Others went full blast to design a machine that, without any external energy source, would produce enough power to keep on going forever. Many were men of religious conviction, and a small number were threatened with ex-communication for such heretical[5] experiments. But they believed that if nature could exhibit perpetual movements – every morning the sun rose, every evening it set; twice a day the tides ebbed[6] and flowed – then man could surely emulate[7] it.

他宣布了结果，王冠中确实掺有白银。后来他把排水定律谦虚地称为“阿基米德定律”。

而阿基米德的另一条妙计，则是发明了一种螺旋泵，帮助希腊建造了海军历史上截至当时最大的军舰“锡拉库西亚号”。当时希腊经典造船工艺还不太合格，军舰每次航行都会进水，这点太让人难以放心了。阿基米德的螺旋泵解决了这个问题。通过圆筒内的一个旋转螺丝，就可以把水不停地汲出：水推动螺丝，螺丝推动泵，几乎通过自身的动力就把流进来的水排出去了。更妙的是，螺旋泵还可以使水向上流。总之，成果非凡。

在之后的几个世纪里，科学家不断探索改进工作原理，希望能使机器不借外力运行。如果机器可以不借外力运行，而且可以永无止境地不借外力运行，那么人类就可以高枕无忧，让机器保持世界的运转。从那时起，对永恒动力的追求使无数发明家发疯、破产或自杀。直到有人发现一种打破两条基本物理定律的方法之前，永动机就永远不会实现。好消息是，打破基本物理定律正是科学家所热衷的挑战。

具有创新精神的发明家不断地研究着永动机这一概念，其中包括发明了珍妮纺纱机的理查德·阿克莱特（Richard Arkwright）以及铁路之父乔治·史蒂文森（George Stephenson）。其他人则竭尽全力设计一种无需外界能量即可产生足够动力永远运转的机器。大部分人是出于宗教信仰，其中少数人则因从事这种异端实验而遭到了被逐出教会的威胁。但他们都相信如果自然界存在永恒运动——太阳晨起夕落，一日潮涨潮落——那么人类完全

While the first kind of water motor was the Buddhist praying wheel, in which prayers are fastened to a wheel and rotated using water power, serious attempts to keep a machine in action in perpetuity① didn't emerge until the Renaissance – and not before Bruges mathematician Simon Stevinus tried to put a dampener on the whole enterprise by tying fourteen balls to an endless piece of cord, suspending them from a triangular frame, and through the law of equilibrium② proving that perpetual motion could not exist. Shortly after, at the beginning of the seventeenth century, Florentine③ mathematician Galileo Galilei fastened one end of a cord to a nail in a wall and swung a heavy ball from it. Although he almost inadvertently invented swingball, he unfortunately couldn't keep the exercise going for long. Perpetual motion was proving evasive.

It fell to an English man to develop the concept further. In 1618 aristocratic courtier Robert Fludd developed a watermill than didn't need a flowing river to run it. The idea was to provide the miller with all the power he needed to grind④ grain, all day, every day, forever. But Fludd was a philosopher and religious fundamentalist, as well as a doctor, alchemist⑤ and inventor, so first he had to define what 'forever' meant. This was not simple. By definition, perpetual motion machines must move in perpetuity. But does that mean for a few years until it breaks down, until God calls time on man, which was the prevailing Christian view, or until the universe caves in on itself, which hadn't even been thought of ? Time is a tricky physical and philosophical concept, and designing a machine that would move till its end made the brains of even the smartest inventors throb.

All of this concerned Fludd, a very serious Jacobean thinker. Often insightful, frequently a fantasist, he was, in many respects, dangerously ahead of his time: claiming, like Galileo,

注释

① perpetuity [ˌpɜːpɪ'tjuːɪtɪ] *n.* 永久
② equilibrium [ˌiːkwɪ'lɪbrɪəm] *n.* 平衡
③ florentine ['flɒrəntain] *adj.* 意大利佛罗伦萨的，佛罗伦萨画派的
④ grind [graɪnd] *v.* 把……磨碎
⑤ alchemist ['ælkəmɪst] *n.* 炼金术士

可以效法自然。

尽管第一种水动机是佛教的祈祷轮（信徒被固定在转盘上，然后利用水力旋转），不过保持机器永恒运动的真正尝试直到文艺复兴之后方才出现，但不会晚于布鲁日数学家西蒙·斯蒂文（Simon Stevinus）证明永恒运动不可能存在的那个实验，因为他的实验让永动机事业彻底变成了不可能。他把14个等重的小球均匀地用线穿起，组成首尾相连的球链，挂在三角形框架上，然后通过守恒定律证明了永恒运动不可能存在。不久之后，17世纪初，佛罗伦萨数学家伽利略·伽利雷（Galileo Galilei）把绳子的一端系在墙上的钉子上，在另一端系上一颗重球。虽然他无意中发明了摆球，但不幸的是他无法保持摆球长时间运动。这就证明了永恒运动是难以实现的托辞。

▼ *17世纪罗伯特·弗拉德设计的永动机的示意图。*

A seventeenth-century illustration of Robert Fludd's perpetual motion machine.

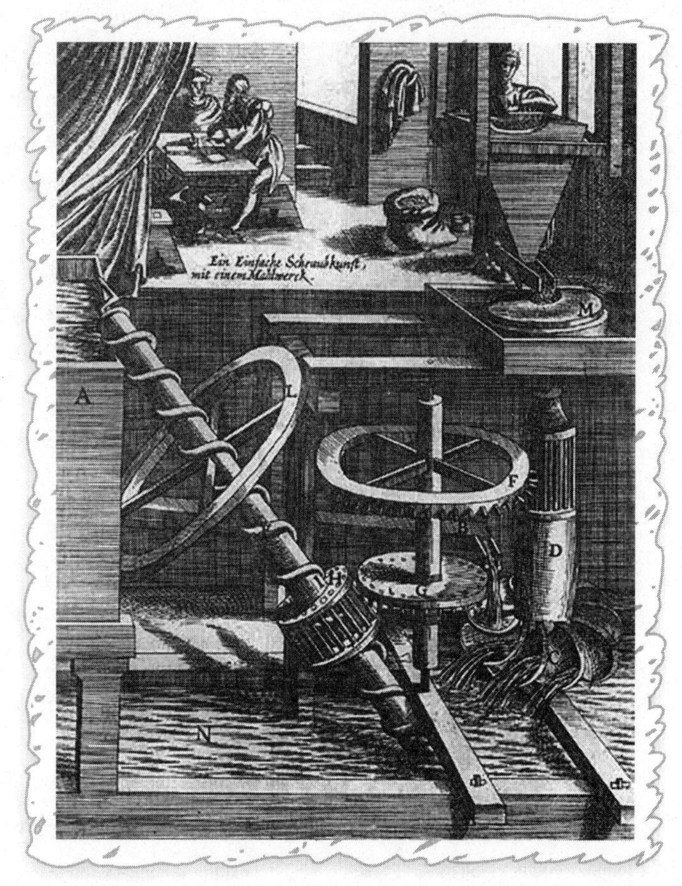

一位英国人进一步发展了这一概念。1618年，贵族的食客罗伯特·弗拉德（Robert Fludd）研制了一种不需要流水推动的水磨。他的想法是日复一日地永远为碾磨工提供碾磨谷物所需的全部能量。但弗拉德是一位哲学家和宗教原教旨主义者，而且还是医生、炼金师和发明家，所以他首先得定义“永远”的意思。这可不简单。从定义来看，永动机必须永远运

注释

① haul [hɔːl] *v.*（用力地）拉
② divinity [dɪ'vɪnɪtɪ] *n.* 神学
③ infiltrate ['ɪnfɪlˌtreɪt] *v.* 渗入，潜入（某地方或组织）
④ betide [bɪ'taɪd] *v.*（某人）倒霉
⑤ apothecary [ə'pɒθɪkərɪ] *n.* 药剂师
⑥ pacify ['pæsɪˌfaɪ] *v.* 安抚
⑦ abiding [ə'baɪdɪŋ] *adj.* 持久的（感情、记忆、兴趣）
⑧ exquisite [ɪk'skwɪzɪt] *adj.* 精美的

that the sun, not the earth, was the centre of the universe – a theory that had the Italian, but not Fludd, hauled[①] up in front of the Pope. He was also convinced that blood moved in a circular motion around the body, something not broadly accepted in seventeenth-century England. Divinity[②] informed much of Fludd's philosophy too. Lightning was the will of God, he insisted, and those who got struck by it got everything they deserved. His theory of disease was somewhat unusual for a man of medicine: a wind controlled by a powerful evil spirit excites lesser evil spirits in the air which then infiltrate[③] the body – and if you don't get to the apothecary fast – ideally the one he ran from his home in London's Fenchurch Street – woe will almost certainly betide[④] you.

With so much going on in Fludd's mind, what with the apothecary[⑤] to run, the king to pacify[⑥] occasionally and learned books to write, he sometimes found little time to devote to his lifetime's abiding[⑦] interest: wheat. But whenever he could, he experimented with the cereal, writing extensively on what he called 'The Excellency' of the crop. His studious interest in milling turned into a series of exquisite[⑧] drawings of a closed-cycle mill using a form of Archimedes' Screw. Water turned a wheel to power a pump that would cause the water to flow back over the wheel. This would then power the pump to cause the water to flow back over the wheel, and so on and so on until everyone got dizzy. And although all this detail was only on paper, the physical laws that would invalidate the concepts were still a long way from being understood.

Although he wasn't the first to design a perpetual motion machine, Fludd was the inventor of the closed-cycle mill. In the following decade, Italian philosopher and alchemist Mark Antony Zimara designed a perpetual motion machine in the form of a self-blowing windmill, in which two or more bellows

行。但“永远”这个词是意味着在它坏掉之前可以运行几年，还是一直运行到上帝消灭人类（这是当时盛行的基督教观点），还是持续到宇宙自我毁灭（这是前人没有想过的情况）？时间是个棘手的物理概念和哲学概念，而设计一台永远运行的机器，即使让最聪明的发明家干，也会计穷力竭。

这一切都是弗拉德所关心的，他是一位非常严肃的詹姆士一世时代的思想家。他在众多领域往往见解独到，富有想象力，想法超前：他像伽利略一样宣称太阳是宇宙的中心，而地球不是，这是一个使意大利——而不是他本人——在教皇面前受审的理论。他还认为血液在人体内循环运动，这在17世纪的英格兰还未被人们广泛接受。神学也极大地影响了弗拉德的哲学思想。他坚信雷电是上帝的意志，而遭到雷击的人罪有应得。他关于疾病的理论对于身为医生的人来说也是不同寻常的：受强大幽灵控制的一股风刺激了空气中较弱的幽灵，然后渗透进身体。如果你不能赶紧看医生（在理想情况下是去他在伦敦芬丘奇街自家开的药店），那么噩运必然会降临在你身上。

弗拉德的头脑中有这么多东西，一会儿忙药店的事务，一会儿要安抚国王，一会儿又要写书，他有时发现自己很少有时间献给终身持久的兴趣——小麦。不过只要他有时间，就用谷物做实验，广泛地写关于这种作物堪称“精华”的文章。他对碾磨的浓厚兴趣转变成了一系列采用阿基米德螺旋泵的闭环磨粉机的精细图纸。水推动轮子转动，轮子给泵提供动力，再引水流回轮子。这样又给泵提供动力，引水流回轮子，如此往复，直到所有人都晕

puff air produced out of nowhere at the arms of the windmill. He was a little vague about how to produce the energy, leaving it 'to the ingenuity of the maker', but was convinced that the arms would rotate forever. By 1653 Edward Somerset, the Marquis of Worcester, who some believe created the first practical steam engine, had a plan 'to raise water constantly, with two buckets[①] only, day and night, without any force but its own motion'. After several long days and nights he eventually concluded it wouldn't work. Over in Germany, Ulrich von Cranch's 1664 machine dropped a ball on top of a paddle-wheel which turns the wheel as the ball falls, rotating an Archimedean Screw. This then carries the ball back to the top of the device when it reaches the bottom. Then the ball falls and the process starts all over again. For the purposes of efficiency, von Cranach used more than one ball and spent hours calibrating the device. But it never overcame the laws of physics to keep going without someone dropping a ball now and again. And in 1686, Georg Andreas Böckler produced numerous ideas for perpetual-motion machines, including 'self-acting mills', in which cups or buckets on an endless rope re-deliver water to a wheel, rotating the wheel and moving the cups in a virtuous[②] loop. It worked perfectly on paper, but nowhere else.

The natural laws of thermodynamics[③], still almost 200 years away from being defined, prevented the seventeenth-century perpetual-motion machines – and all subsequent and, in all probability, future ones – from working. The first and second laws of thermodynamics say, respectively, that energy can be neither created nor destroyed but only changed in form, and that it is impossible to make a machine that doesn't waste some energy, however

注释

① bucket ['bʌkɪt] *n.*（有提梁的）桶
② virtuous ['vɜːtʃʊəs] *adj.* 品德高尚的
③ thermodynamics [ˌθɜːməʊdaɪ'næmɪks] *n.* 热动力学

了。虽然所有细节仅见于纸上，但要想理解证明这个想法无效的物理定律还得走很长的路。

虽然弗拉德不是设计永动机的第一人，但他是闭环磨粉机的发明人。10年之后，意大利哲学家和炼金师马克·安东尼·兹马拉（Mark Antony Zimara）设计了一台自吹风风车式永动机，风车臂可以无中生有地刮出两股以上的风。他对如何产生能量还有点儿茫然，就把这留给了“造物者的巧夺天工”，不过他还是确信风车臂会永远转下去。到了1653年，伍斯特侯爵爱德华·索默塞（Edward Somerset）（有人相信他创造了第一台实用的蒸汽机）有了一项“只靠两只桶自身运动而不靠外力就可以昼夜不停汲水”的计划。经过夜以继日的工作，最后他得出结论：这台机器无法运行。而在德国，乌尔里希·冯·克兰奇（Ulrich von Cranch）在1664年设计了一台机器，把一枚圆球从叶轮顶部掷下，当圆球落下时转动叶轮，从而旋转阿基米德螺旋泵。当螺旋泵旋转到底部时，又把圆球送到机器的顶部。接着圆球再次落下，整个过程又重新开始。为了提高效率，冯·克兰奇用了多枚圆球，而且花了大量时间校准机器，但最终还是未能克服物理定律，如果没人重复掷下圆球，机器就无法保持运转。1686年，乔治·安德烈亚斯·伯克勒（Georg Andreas Böckler）想出了无数的永动机创意，包括“自动磨粉机”。在这种磨粉机中，系在首尾相接绳子上的水杯或水桶向轮子上送水，使轮子旋转起来，带动水杯做环形运动。不过也就是纸上谈兵而已。

这时离热力学自然定律的提出还有200多年，这条定律证明了17世纪以及后来的所有永动机不可能运行。热力

注释

① friction ['frɪkʃən] *n.*（人与人之间的）摩擦

small. In practice, the second law of thermodynamics means that any friction[①] created by the wheel of a perpetual motion machine **turns into heat and noise, thus losing energy**. And if the machine **for whatever reason doesn't cause friction**, it will still need more energy than it produces to keep the wheel itself going, thus breaking the first law. In short, perpetual motion machines are impossible, yesterday, today, and although one should never say never, in all likelihood forever. The nearest we get today are nodding dogs in the backs of cars, and only then when the vehicle is in motion.

学第一定律和第二定律分别指出，能量既不能产生也不能消亡，而只能转换形式；不可能制造出不消耗能量的机器，无论多小也不可能。在实践中，热力学第二定律意味着永动机轮子产生的摩擦转换为热量和噪声，从而消耗了能量。如果永动机因为某种原因不产生摩擦，它仍然需要比所产生能量更多的能量保持轮子自身运转，从而打破了热力学第一定律。总之，永动机不可能被造出来，昨天不可能，今天不可能，很可能永远也不可能——虽然永远也不应该说“永远”。而我们今天与永动机最接近的东西就是汽车后座的“摇头狗”，但它也只是在汽车运动时才会动。

REJECTED

BENTHAM'S ALL-SEEING PANOPTICON

It's Big Brother's Big Brother. More than 150 years before George Orwell, philosopher Jeremy Bentham designed the panopticon; a building that would be 'all-seeing' and where people, fearing observation even when no one was present, would feel compelled[①] to behave. Simply designed, ruthlessly efficient and although almost wholly ignored, not least by King George Ⅲ who pulled funding from the first one, it was said that panopticon buildings would do good for society, good for the people being observed and good for the public purse.

Reflecting Bentham's philosophy, the essence was this: panopticon buildings made people behave through their simple but practical design: a circular or semi-circular block around a central base, generally a tower, from which surveillance[②] could be conducted – or not. Every individual cell, indeed every individual, could be under watch at any time of night and day. Quite possibly – and here is the panopticon's genius – in reality no one would be monitoring very much at all. Everyone who was forced to reside within the building for whatever reason would see the tower, but not be able to penetrate[③] its windows to see what lay beyond. Warders may be there, keeping watch,

注释

① compel [kəm'pɛl] *v.* 迫使

② surveillance [sɜː'veɪləns] *n.* 监视

③ penetrate ['pɛnɪˌtreɪt] *v.* 进入，穿透

项目遭拒

边沁的全视圆形监狱

在作家乔治·奥威尔（George Orwell）之前150多年，哲学家杰里米·边沁（Jeremy Bentham）设计了一种圆形监狱，它是老大哥中的老大哥，是一种“全视”建筑，就算没人监视，里面的人也会因为害怕有人在观察他们，而被迫注意自己的行为。圆形监狱设计简单、无情而高效，尽管几乎被完全忽视，尤其是乔治三世国王撤回了提供给第一座这种监狱的资金，不过据说这种监狱有益于社会，有益于被监视的人，有益于国库。

如果对边沁的学说进行深思的话，它的精髓是这样的：圆形监狱使人约束自己的行为，是通过简单而实用的设计实现的——四周环形或半环形的囚室围绕着中心，中心通常是一座塔楼，从塔楼可以监视囚室。每一间囚室——甚至每个人——昼夜任何时间都可以受到监视。不过，很可能根本没人在那里死盯着监视，但这正是圆形监狱的高明之处。囚禁在这里的人都可以看到中央的塔楼，但无法透过塔楼窗户看到塔楼里面的情形。看守可能在那

注释

① omnipresent [ˌɒmnɪ'prɛzənt] *adj.* 无所不在的
② gaoler ['dʒeɪlə] *n.* 监狱看守，监狱长
③ asylum [ə'saɪləm] *n.*（给予通常因政治原因不能回国的人的）避难许可
④ manifesto [ˌmænɪ'fɛstəʊ] *n.* 宣言
⑤ quaint [kweɪnt] *adj.* 古雅的
⑥ circumference [sə'kʌmfərəns] *n.* 周长

ready to enforce order, happy to punish. But equally they may be absent, and it is this concept of the uncertain, that Big Brother could be watching you, rather than definitely is, that gave the panopticon its power of control.

Children, at desks in isolated cells, would behave exceptionally and attend to their studies dutifully in panopticon schools, where an omnipresent[①] teacher could descend on them whenever attention wandered. Prisoners would conform to their gaolers[②]' rules when not at work on the treadmill or spinning looms, which Bentham, one of the eras great liberal thinkers, believed would be productive activities. The insane would expect to be under constant watch in their panopticon hospital or asylum[③]. Any attempt to flee would be madness. For the panopticon was essentially a building of mind control. Bentham called it 'the inspection house … a method of controlling mind over mind'. While surveillance was a possibility rather than a certainty, pupils, prisoners or patients would indulge in self-surveillance. The belief that one is under constant watch can do much for the mind.

Bentham's panopticon manifesto[④] was written in quaint[⑤], polite terms. 'The building is circular,' he explained.

> 'The apartments of the prisoners occupy the circumference[⑥]. You may call them, if you please, the cells. The apartment of the inspector occupies the centre; you may call it, if you please, the inspector's lodge.' In their cells, inmates could see little but the tower. Even shadows out were designed to prevent inspectors revealing their presence. Each cell ran for the width of the building with an inside window looking out on the tower, and an outside one allowing light to pass through. Instead of doors, simple openings spaced so no one

里监视，准备维持秩序和惩罚犯人，但同样他们也可能不在那里。正是这种不确定的概念，即看守可能在监视你，而不是一定在监视你，使圆形监狱有了控制的力量。

孩子们在独立的房间里上课，在圆形学校里他们异乎寻常地听话，学习也相当认真，在那里只要他们走神就会有无处不在的老师突然来到他们面前。作为那个时代伟大的自由主义思想者之一，边沁认为让囚犯踏踏板或在织布机上干活是生产性的活动，而当囚犯不在踏板或织布机干活的时候，则应该服从看守的支配。精神有问题的人会认为自己在全视医院或救济院中被持续看守，任何逃跑的尝试都是愚蠢的行为。因为这种圆形的设计从本质上来说是控制思想的建筑。边沁称“监视房……是一种用思想控制思想的方法”。由于监视只是一种可能而不是一定，学生、囚犯或病人就会时刻处于自我监禁的状态。认为自己一直被监视的想法对这种自我监禁的思想起了很大作用。

边沁对圆形监狱的声明措辞精巧而文雅。他解释道：

> 建筑是圆形的，整个一圈都是囚犯的住所，如果你愿意的话，可称之为“小室”。中央是看守的住所；如果你愿意的话，可称之为“看守的小屋”。小室里的犯人除了塔楼什么都看不到。连外面遮盖物的设计都是为了避免犯人发现看守。每间小室与建筑等宽，朝向中心的窗户面对塔楼，朝外的窗户用来通光。小室没有门，只有按照规定距离排列的开口，这样囚犯们互相看不到，既能避免犯人发出嘈杂声，也能避免万一光

inmate could see another eradicated[①] both noise and the chance of light from a half-opened door betraying the presence of a warder[②] or teacher. 'All play, all chattering – in short, all distraction of every kind – is effectually banished[③] by the central and covered situation of the master, seconded by partitions[④] or screens between the scholars, as slight as you please,' wrote Bentham. 'That species of fraud[⑤] at Westminster called cribbing, a vice thought hitherto congenial[⑥] to schools, will never creep in here.'

By design, through fear, the panopticon would 'introduce tyranny[⑦] into the abodes[⑧] of innocence and youth,' which, although not easily reconciling[⑨] with his liberal tendencies, Bentham thought a jolly good thing. Conversely, the panopticon prison would put an end to random acts of cruelty by warders. There was a financial advantage too. Because inmates and schoolboys would control their own behaviour, panopticons required far fewer staff and thus the wage bill could be slashed[⑩]. Bentham's death in 1832 was just two years before new Poor Laws heralded[⑪] the construction of workhouses, when millions, destitute[⑫] and desperate, threw themselves on the mercy of their parish[⑬]. In return for food, accommodation and ritual humiliation and cruelty, the state would expect absolute control on every aspect of their lives. The panopticon workhouse would be perfect. Many nineteenth century institutions – workhouses, prisons, hospitals – owed much to Bentham's design, if not his philosophy, which was to provide: 'the greatest happiness for the greatest number of people'. In many respects, he fell short on this one.

While plenty of buildings owed a nod to panopticon design, in reality almost no truly authentic such buildings were

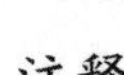

注释

① eradicate [ɪ'rædɪˌkeɪt] *v.* 根除
② warder ['wɔːdə] *n.* 看守
③ banish ['bænɪʃ] *v.* 驱逐
④ partition [pɑː'tɪʃən] *n.* 隔断
⑤ fraud [frɔːd] *n.* 诈骗罪
⑥ congenial [kən'dʒiːnjəl] *adj.* 令人愉快的
⑦ tyranny ['tɪrənɪ] *n.* 专制暴政
⑧ abode [ə'bəʊd] *n.* 住所
⑨ reconcile ['rɛkənˌsaɪl] *v.* 使和谐一致，调和
⑩ slash [slæʃ] *v.* 砍，劈
⑪ herald ['hɛrəld] *v.* 预示……的来临
⑫ destitute ['dɛstɪˌtjuːt] *adj.* 赤贫的
⑬ parish ['pærɪʃ] *n.* 教区

线从半开的门穿过，让囚犯发现看守或者教师在监视。所有玩乐，所有闲谈——总之每种分散注意力的消遣——都被在中央隐蔽着的教师有效禁止了，另外学者们之间也有隔断或屏风，你可以把它们尽量做得不显眼。在议会里有种欺骗类型被称作剽窃，这种恶劣思想至今在学校里还很常见，却永远不会出现在这里。

圆形学校的设计，让人担心会“把专制带入无辜的人和青年人的住所”，这虽然不太与边沁的自由主义倾向一致，但他仍然认为这是一件极好的事。相反，圆形监狱会结束看守任意的残暴行为。也有经济利益，因为囚犯和学生将控制自己的行为，所以圆形监狱或学校需要的工作人员便更少一些，因此需要支付的工资总额也会减少。边沁于1832年去世，正值新《济贫法》颁布的两年之前，这时成百上千万贫穷绝望的人正指望着教区的仁慈新法颁布后建立济贫院。国家为这些人提供食物、住宿，对他们照例进行残酷的羞辱，希望能以此换来对他们生活的各方面进行绝对的控制。全视救济院将是完美的选择。19世纪的很多机构——救济院、监狱、医院——如果不是归功于边沁的学说“为最多的人提供最大的幸福”，也要在很大程度上归功于他的设计。他在很多方面都未能履行他的学说。

虽然很多建筑要用全视设计，但事实上几乎没有几座是按这种设计建成的，边沁活着的时候肯定没有。这令人遗憾，当然代价也高。边沁虽然很富有，但也在开发圆形监狱的过程中花费了巨资，最终说服了政府给他提供一个场地。

constructed, certainly not in Bentham's lifetime. This was unfortunate – and costly too. Although very wealthy, Bentham spent a large sum of money developing the panopticon prison, eventually persuading the government to provide a location. His plans were scuppered when George Ⅲ refused to authorise purchase of the site. In 1813, Bentham received £23,000 to go some way towards making good his loss. It was small recompense for his efforts. If only the king had dug deeper, institutions such as the workhouse might have been creepier[①], but somewhat more humane.

Despite his losses on the panopticon prisons, Bentham left a sizable legacy[②] to the University of London – what today is University College, London – and to Edwin Chadwick, the creator of the Pure Air Company. Ending his life an absolute rationalist, on his own request and to show that death really is the end and resurrection a myth, his body was dissected, embalmed, re-dressed in his own clothes – large hat, cutaway coat, nankeen trousers – and propped in a chair where it can be seen to this day in a display case at the university's Gower Street base. Bentham's head, sadly, is not his own, after the embalming[③] process went wrong, leaving his face disfigured. Instead he now has a wax one. The real head, having survived safely in the college for nearly 150 years, was stolen by students in 1975 and held for ransom[④]. After going missing several times subsequently, it is now kept in storage at an unknown location. His legacy, though, lives on; so to that extent, for Bentham at least, there is an afterlife. Benthanism gained a significant philosophical following that was both radically liberal and ultra-controlling ('bossy Benthamites', historian A.N. Wilson calls them).

Today panopticism is not much more than a concept in criminology[⑤] and social science; brought back into fashion

注释

① creepy ['kri:pɪ] *adj.* 吓人的
② legacy ['lɛgəsɪ] *n.* 遗产
③ embalm [ɪm'bɑ:m] *v.* 对（尸体）进行防腐处理
④ ransom ['rænsəm] *n.* 赎金
⑤ criminology [ˌkrɪmɪ'nɒlədʒɪ] *n.* 犯罪学

但乔治三世不批准购买他的这处场所，边沁的计划也就落空了。1813年，边沁收到2.3万英镑，以补偿他的部分损失。相对于他的努力来说，这点补偿太少了。若是国王当时再进一步，济贫院这样的机构可能会更加“吓人”，但也多了几分人道。

尽管边沁在圆形监狱上遭受了损失，但他还是给伦敦大学（现在的伦敦大学学院）和纯净空气公司的创建者埃德温·查德威克（Edwin Chadwick）留下了大笔遗产。他终其一生都是绝对的理性主义者，应他要求为了表明死亡是真正的终结，复活是一种神话，他的尸体被解剖，做了防腐处理，重新穿上他自己的衣服——大帽子、常礼服和淡黄色裤子——被架在椅子里。至今都能在大学的高尔街基地的陈列箱里看到他。不幸的是防腐过程出现问题，损坏了他的面部，所以现在尸体的头部不是边沁自己的，是用蜡像代替的。他真正的头部在学院安全保存了将近150年，1975年被学生偷走，用来勒索赎金。后来又丢了几次之后，现在被保存在一处不为人知的地方。但他的遗产还在，从这个意义上来说，边沁至少还有来世。边沁主义学说获得了一批重要的追随者，他们既是激进的自由主义者又是极端的控制者（史学家威尔逊称他们为“专横的边沁学派”）。

现在，边沁主义仅仅是犯罪学和社会科学的一个概念了。法国哲学家米歇尔·福柯

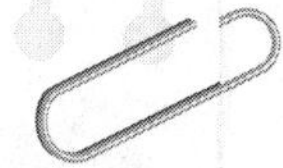

▼ 英国伦敦大学学院里放置的改革家杰里米·边沁的遗体（头是蜡像）。

Reformer Jeremy Bentham's cadaver (with wax head) at University College London.

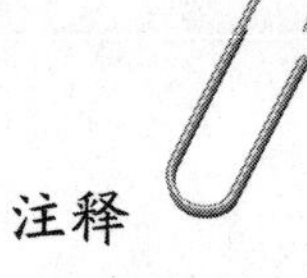

注释

① unverifiable [ʌn'verɪfaɪəb(ə)l] *adj*. 无法核实的，无法检验的，不能证实的

by French philosopher Michel Foucault whose 1975 work *Discipline & Punish: The Birth of the Prison* is very widely read in very small circles. 'The major effect of the Panopticon [is] to induce in the inmate a state of conscious and permanent visibility that assures the automatic functioning of power,' writes Foucault. 'Bentham laid down the principle that power should be visible and unverifiable[①].' And the concept is retained in cities and towns throughout the world through the use of CCTV cameras. We can all be seen – possibly – whatever we do, wherever we go. Panopticism survives, even if the buildings never made it.

（Michel Foucault ）1975年的著作《纪律与惩罚：监狱的诞生》在很小的圈子里被广为阅读，是他使边沁主义重新流行起来。福柯写道："圆形监狱的主要作用是诱导囚犯处于自觉状态，让他们以为自己一直被监视着，这样能保证权力自动发挥效用。边沁认为权力应该是可见的且无法证实的。"而这个概念在全世界的城镇中，以闭路电视摄像头的形式被保留了下来。不论我们做什么，不论去哪里，我们都会（或可能会）被看到。全视主义虽然没有被用到建筑中，但它却存留了下来。

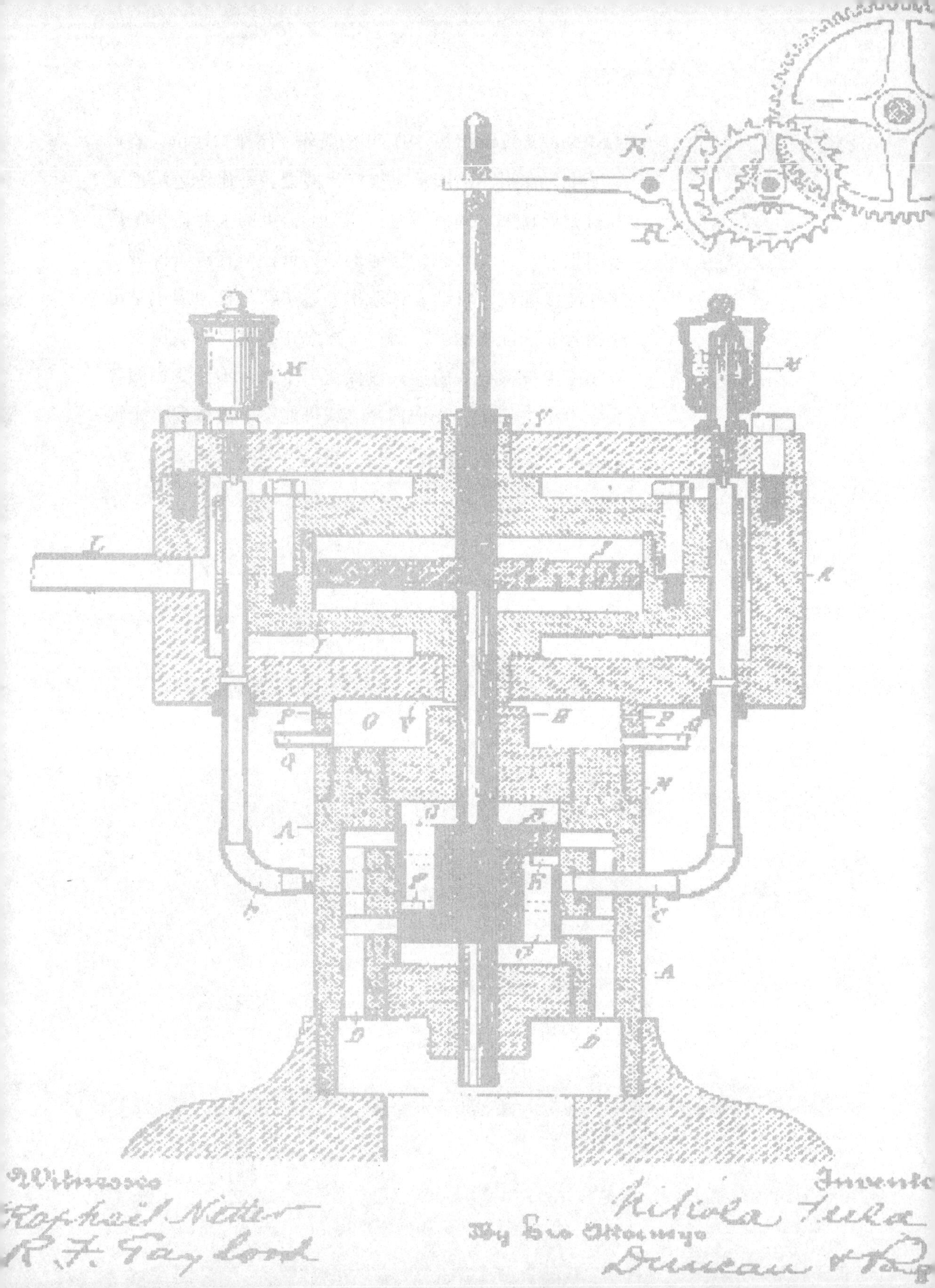
Witnesses
Raphael Netter
R. F. Gaylord
Nikola Tesla
By his Attorneys

PART TWO

第二部分

TESLA'S EARTHQUAKE MACHINE

Famous in the early part of the twentieth century, Nikola Tesla deserves to be afforded more honour today. Arguably the inventor of radio (despite Marconi pipping him to the patent) as well as the developer of the alternating current (AC) – which pitted him against electricity giant Thomas Edison – he was the underdog in many battles. However, no one beats Tesla's supremacy in earthquake machines.

His pocket-sized device, the reciprocating[①] mechanical oscillator[②], worked on the principle that every object has a 'resonance[③]' – a point at which it will begin to shake or shatter[④] objects around it – very much like glass does in the presence of a piercing opera singer. With his experiments on high frequencies breaking new ground, such a machine could serve very practical purposes, such as discovering new oil fields or sites of archaeological[⑤] interest.

By 1897, and now a big name in the field of electromagnetic[⑥]

注释

① reciprocate [rɪ'sɪprə,keɪt] *v.*（术语）沿直线往复移动
② oscillator ['ɒsɪ,leɪtə] *n.* 振荡器
③ resonance ['rɛzənəns] *n.* 共鸣
④ shatter ['ʃætə] *v.* 粉碎
⑤ archaeological [,ɑrkɪə'lɑdʒɪkl] *adj.* [古] 考古学的
⑥ electromagnetic [ɪ,lɛktrəʊmæg'nɛtɪk] *adj.* 电磁的

特斯拉的地震机

尼古拉·特斯拉（Nikola Tesla）在20世纪早期就声名卓著，但他现在应该得到更高的荣誉。毫无疑问，他发明了无线电，但专利权被意大利物理学家马可尼（Marconi）夺走。还发明了交流电——与电力研究巨人托马斯·爱迪生（Thomas Edison）对阵——但他在多次斗争中都败下阵来。然而，没人能在地震机领域超越他的权威地位。

他发明的往复式机械振荡器只有口袋大小，工作原理是一切物体都有共振点，意思是物体只要达到这个点，就开始振动或震碎周围的物体，很像歌剧演员唱高音时会把玻璃震碎的情景。他对高频的实验有了新突破后，这种仪器就可以有很实际的用途，例如用于发现新油田或者古迹遗址。

1897年，特斯拉成为电磁领域的重要

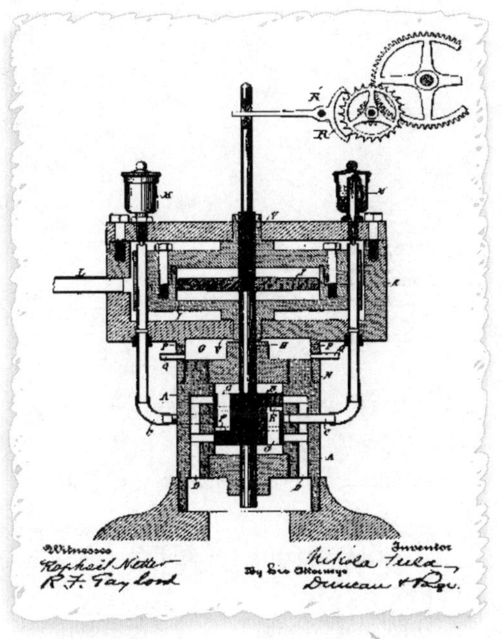

▲ 尼古拉·特斯拉的往复式机械振荡器或地震机的专利图。他称这台仪器可用于探测新油田。

A patent drawing for Nikola Tesla's reciprocating mechanical oscillator or 'earthquake machine', which he claimed could be used to discover new oil fields.

discovery, Tesla had already shown how electrical energy could be transmitted[①] wirelessly – something that is today known as the Tesla effect. He'd partnered with American electric company owner George Westinghouse to promote alternating current, and was working in a central Manhattan laboratory on new projects. Amongst them, he wanted to produce a device that, by resonating in tune with the natural vibrations[②] that all structures make, could shake buildings to their core. It wasn't long before he had a small machine that was up to the job of pulling things down.

Forever treading a fine line between sanity[③] and madness, as Tesla turned on the machine in his lab one day he felt the walls shudder. Excitement got the better of him and he ratcheted[④] up the power. 'Suddenly, all the heavy machinery in the place was flying around,' he claimed. 'I grabbed[⑤] a hammer and broke the machine. Outside in the street, there was pandemonium[⑥].' New York's emergency services swung into action but could find no cause of the trouble. Tesla swore his assistants to secrecy, but the word on the street was that an earthquake had struck. And the cause: the Tesla oscillator, earthquake machine.

The story of Tesla's shaking lab has never been proven, but the inventor, eager to register a patent ahead of the rivals who had humiliated[⑦] him in the past, was happy to embellish[⑧] it, thus talking up the potential of his small, light, steam-powered device regardless. When suitably tuned, Tesla claimed, the oscillator could destroy anything. Resonance was already known to damage buildings and bridges. Now here was a gadget[⑨] that pushed the laws of physics to their limits and beyond. 'Five pounds

注释

① transmit [trænz'mɪt] *v.* 传送，输送，发射

② vibrations [vaɪ'breɪʃənz] *n.* 心灵感应，共鸣

③ sanity ['sænɪtɪ] *n.* 心智健全，神志正常

④ ratchet ['rætʃɪt] *v.*（使）逐渐小幅增长

⑤ grab [græb] *v.* 抓住

⑥ pandemonium [ˌpændɪ'məʊnɪəm] *n.* 喧嚣，嘈杂

⑦ humiliate [hjuː'mɪlɪˌeɪt] *v.* 使蒙羞

⑧ embellish [ɪm'bɛlɪʃ] *v.* 装饰

⑨ gadget ['gædʒɪt] *n.* 小器械（有时暗指复杂、不必要的东西）

人物。他已经展示了如何通过无线传播电能——现在被称作特斯拉效应。他与美国西屋公司负责人乔治·西屋（George Westinghouse）建立合作关系，共同推广交流电。他还在曼哈顿中心实验室研究一些新项目。在这些新项目之中，他想制造一种设备，能和一切建筑的自然振动发生共振，从而使建筑物从表到里都产生振动。不久之后，他就发明了一台小机器，可以把一些东西推倒。

特斯拉始终在聪明和疯狂之间打擦边球。有一天，他在实验室打开机器后感觉墙开始振动。他异常兴奋，慢慢调高功率。他说："突然，在场的所有重机器都飞了起来。我拿锤子把那台机器砸烂。外面街上一片混乱。"纽约的应急服务人员飞速赶往现场，但没能找到事故起因。特斯拉要求助手保守秘密，但街上传言说刚才发生了地震，而起因是特斯拉振荡器，也就是地震机。

特斯拉实验室振动的事情从未得到证实，但这位发明家急于抢在曾经羞辱他的对手之前注册专利，于是乐此不疲地对这件事添枝加叶，大谈他的设备不仅小巧玲珑而且采用蒸汽动力，具有很大潜力。他说，只要调谐适当，这台振荡器就可以摧毁任何东西。人们已经知道共振可以毁坏建筑物和桥梁，现在，这个小玩意儿把物理定律推到了极限甚至更远。"5磅气压……我只需要这么大的力就能把帝国大厦推倒，"他说，"用同一台振荡器，我还可以用不到1小时的时间摧毁布鲁克林大桥。"

这项技术相当简单。当把蒸汽注入振荡器时，这台设备就开始振动，直到达到自然共振，然后就试图把自己震开。把这台设备连接到建筑物上，然后振荡器的每次振动

注释

① vibrate [vaɪ'breɪt] *v.* 使颤动，颤动
② empirical [ɛm'pɪrɪkəl] *adj.* 实证的
③ scaffolding ['skæfəldɪŋ] *n.* 脚手架
④ beam [biːm] *n.*（建筑物的）梁

of air pressure … that is all the force I would need' to fell the Empire State Building, he said. 'With the same oscillator I could drop the Brooklyn Bridge in less than an hour.'

The technology was relatively simple. When steam is forced into the oscillator, the device vibrates[①] until it reaches its natural resonance when shocks begin and the device then tries to shake itself apart. Attached to a building, each beat of the oscillator can be made to link to the structure's own natural vibrations. With each stroke, the force is magnified, and once the frequency of the resonance equals the time taken for vibrations to spread throughout the building, there's trouble. The lower the resonant frequency, the easier it is. Only a small force is needed to generate localised Armageddon, as the resonant effects are magnified and terrifying.

After his claim about a Manhattan earthquake, Tesla sought further empirical[②] data. Strapping the alarm-clock-sized vibrating machine to a 2ft-long metal bar, he hit New York's streets in search of a construction site where half-erected steel frames could be turned into not-at-all erected mashes of mush. Near Wall Street such a building had reached ten-storeys high and, ignoring the workmen on the scaffolding[③], he quickly strapped an oscillator to one of the beams[④]. Tesla's account says that the structure began to crack and that builders came down to ground level in panic. Police were called once more to investigate another earthquake, no doubt puzzled that central Manhattan may have suddenly developed a geographical fault line. Tesla slipped away with the machine in his pocket, delighted that he could have destroyed a steel building within minutes. Or at least, so he said.

While the patent and a model for Tesla's earthquake machine exist, there's actually little evidence to verify his stories. This seemed not to matter to him. His interest in making the earth

都与建筑物本身的自然振动联系起来。随着不断振荡，力逐渐增强，一旦共振频率等于振动传遍整座建筑物所需的时间时，就产生了破坏效应。共振频率越低，越容易产生破坏效应。由于共振效应极大，十分恐怖，所以只需要一点儿力就可以引发局部末日决战。

特斯拉在声称曼哈顿地震之后，进一步收集实验数据。他把闹钟大小的振荡器绑在0.61米（2英尺）长的金属杆上，去纽约街头寻找一处施工场地，想看看能否把那里半立的钢筋框架变成根本立不起来的烂泥。华尔街附近有一座已经盖了10层的大楼，他不顾工人们还在脚手架上施工，迅速把振荡器绑在一根横梁上。据他自己说，大楼开始出现裂缝，工人们十分恐慌，赶紧跑到地面上来。又一次发生“地震”，警察又得来调查，他们一定困惑地以为曼哈顿中央可能突然出现了地质断层线。特斯拉把设备装进口袋里，然后溜掉了。他想到自己能在几分钟内毁掉一座钢结构大楼，非常高兴。至少他是这么说的。

虽然特斯拉的地震机既有专利，也有模型，但几乎没有证据可以证明他的故事。这对他来说似乎并不重要。20世纪初，随着他和托马斯·爱迪生的口水战愈演愈烈，他对制造地动设备逐渐失去了兴趣。爱迪生曾经许诺，如果特斯拉能改进爱迪生的发明（他做到了），将给他5万美元奖金，但爱迪生最后食言了。科学家和新闻媒体对爱迪生怀恨在心，一直在争论特斯拉的交流电和爱迪生的直流电哪个更有优势。直到1903年，交流电在“电流之战”中成为公认标准，特斯拉当然无比高兴。

随着电力的不断普及，爱迪生已经习惯了坐享专利

move diminished as his public spat with Thomas Edison intensified[①] during the first years of the twentieth century. Harbouring a grudge[②] against Edison for the non-payment of a $50 bonus once promised if Tesla could improve upon Edison's inventions (he did), scientists and the press were by now debating the relative merits of Tesla's alternating[③] current and Edison's competing direct current. By 1903, to Tesla's undoubted pleasure, AC had become the accepted standard in the 'war of the currents'.

Having become used to receiving royalty[④] payments as electricity became more commonplace, coming second to Tesla had hit Edison in the pocket. To protect his invention and get back at Tesla, whose increasing oddness could be used to whip[⑤] up concern about his alternating current, Edison decided he needed to convince the public about the inherent dangers of AC. For this purpose, he called upon the service of an elephant awaiting execution. This PR stunt proved quite a shock, not least to Topsy the elephant, and, when AC felled the beast, Tesla's alternating current suffered commercially.

Topsy wasn't the first animal to fall foul of Edison's need to prove his point. Smaller animals had already met their match (being 'Westinghoused' as Edison expressed it, a reference to Tesla's patron[⑥]). Cats, dogs, horses, the odd cow; all were zapped by Edison in private. When news reached him that Luma Park Zoo at Cony Island was top heavy in elephants – Topsy in particular, treading[⑦] keepers to death at the rate of one a year, was proving too much for her handlers – he was delighted. Topsy, the first and indeed only elephantine execution by electricity, may consider herself lucky. The zoo's owners had considered a hanging, but animal protectionists objected on humanitarian[⑧] grounds. Instead, after being fed with carrots laced with cyanide[⑨], and in front of a crowd of 1,500, Edison

注释

① intensify [ɪn'tɛnsɪˌfaɪ] *v.* 加强，强化
② grudge [grʌdʒ] *n.* 积怨
③ alternate ['ɔːltəneɪt] *v.*交替
④ royalty ['rɔɪəltɪ] *n.* 王室
⑤ whip [wɪp] *v.* 鞭打，抽 whip up 煽动
⑥ patron ['peɪtrən] *n.*（艺术家、作家、音乐家等的）资助人
⑦ tread [trɛd] *v.*（小心）行事
⑧ humanitarian [hjuːˌmænɪ'tɛərɪən] *adj.* 人道主义的
⑨ cyanide ['saɪəˌnaɪd] *n.* 氰化物

权转让使用费，但屈居特斯拉之后使爱迪生的收入大幅减少。为了保护发明并击败特斯拉，爱迪生认为有必要使大众认识到交流电的内在危险，而同时特斯拉也变得越来越古怪，这让人们对他发明的交流电变得忧心忡忡。爱迪生为了达到这一目的，找来一头等待处死的大象。这次公开演示的结果非常惊人，不仅击倒了这头名叫托普希的大象，而且由于交流电击倒动物，让特斯拉的交流电蒙受了巨大的商业损失。

托普希并不是第一只被爱迪生为证明自己观点用来满足自己需求的动物。更小的一些动物已经遇到了对手（按爱迪生的说法是“被西屋了”，而西屋正是特斯拉的赞助人）。猫、狗、马、牛都被爱迪生私下里杀死了。他得知康妮岛上月神公园动物园的大象严重超员（尤其是这头名叫托普希的母象，每年踩死一名看守人），非常高兴。托普希作为第一头而且是唯一一头被执行电刑的庞然大物，可能会觉得自己很幸运吧。动物园业主本想采用绞刑，但动物保护主义者出于人道主义反对这样做。因此，喂了托普希拌有氰化物的萝卜后，爱迪生当着1500人的面让6000伏的交流电从大象腿上通过全身。托普希立刻毙命——爱迪生满意了，特斯拉的交流电不安全。媒体如爱迪生所愿报道了这一事件。后来这件事的录像《对大象执行死刑》也在同年公布了（现在还在You Tube网站上保存着，以供后人观看）。

特斯拉最终在电流之战中败下阵来，把他过人的才能转到其他兴趣上，包括燃气动力飞艇（他放弃了当时作为“玩具”的飞机，因为有“致命缺陷”，永远不可能有实

注释

① zap [zæp] *v.*（用枪或在电脑游戏中）杀死，摧毁，击中
② volt [vəʊlt] *n.* 伏特（电压单位）
③ posterity [pɒ'stɛrɪtɪ] *n.* 子孙后代
④ airship ['ɛəˌʃɪp] *n.* 飞船，飞艇
⑤ obsessed [əb'sɛst] *adj.* 受困扰的，对……痴迷的
⑥ napkin ['næpkɪn] *n.* 餐巾，餐巾纸
⑦ spat [spæt] *n.* 口角，小争吵
⑧ deem [diːm] *v.* 认为，相信

zapped① 6,000 volts② of alternating current through the elephant from the legs up. Topsy died instantly – and Edison, satisfied that Tesla's current wasn't safe, got his press coverage. A recording of the event, *Executing an Elephant*, was released later that year (and is now saved for posterity③ on YouTube).

Finally zapped in the war of the currents, Tesla turned his considerable talents to other interests, including gas-powered airships④ (he dismissed the planes then flying as 'toys' that would never become commercially practical because of 'fatal defects'); a new kind of flying machine powered wirelessly from energy stations on the ground; and his own 'death beam'. As life rolled on, his state of mental health diminished along with his fame. He once said: 'One can be quite insane and think deeply.' For a man obsessed⑤ with the number three – he often walked around a block three times before entering a building, demanded three folded napkins⑥ at every meal and would only stay in a hotel room with a number divisible by three – this was deeply apt.

Rather ironically, the only major recognition Tesla received in his lifetime was the Edison Medal; the American electrical engineers' highest award. Marconi had won the Nobel Prize for physics in 1909 for the radio; and Tesla's spat⑦ with Edison meant that both were considered – and overlooked – for the prize in 1915. On his death, the FBI seized Tesla's paperwork, deeming⑧ many of his ideas too secret to release.

际的商业价值）；还包括通过地面能源站无线提供动力的飞行器和他自己发明的死亡射线。随着年龄增长，他的精神状况和他的名气一起衰退。他曾说：“人可以疯狂而同时深思。”对这个对数字3很着迷的人来说——他经常进入一座楼之前绕街区走三遍，每餐都要求餐巾纸折三折，住酒店时只愿意住在房间号能被3整除的房间里——这句话太贴切了。

相当具有讽刺意味的是，特斯拉一生中得到的最重要的荣誉就是爱迪生奖章，这是美国电气工程领域的最高奖章。马可尼于1909年因发明无线电获得诺贝尔物理学奖。1915年特斯拉和爱迪生同时被授予诺贝尔奖，但因两人的口水战，两人都没去领奖。特斯拉一去世，联邦调查局就封存了他写的材料，把他的很多想法当作机密，不予公开。

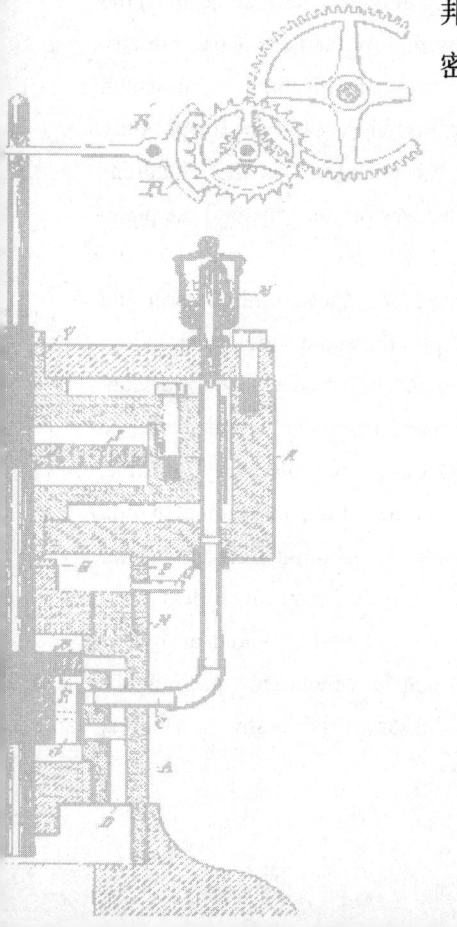

ABANDONED EDISON'S CONCRETE FURNITURE

注释

① execute ['ɛksɪˌkjuːt] *v.* 处死

② congressman ['kɒŋgrɛsmən] *n.*（尤指众议院的）美国国会男议员

③ rattle ['rætəl] *v.* 使发出嘎嘎声，发出嘎嘎声，使紧张

④ henceforth ['hɛns'fɔːθ] *adv.* 从此以后

⑤ repertoire ['rɛpəˌtwɑː] *n.*（表演者的）全部曲目，保留剧目

⑥ philanthropic [ˌfɪlən'θrɒpɪk] *adj.* 慈善的，博爱的

⑦ resonant ['rɛzənənt] *adj.* 洪亮的

Around the same time that he was executing① elephants, Edison developed a plan to murder music. After the failure of his very first invention – the telegraphic vote-reading machine for US Congressmen②, so slow that by the time it had rattled③ into action members had often changed their minds about which way to vote – Edison decided that he would henceforth④ only invent products for which there was a clear demand. Hundreds of ideas later, concrete furniture was on the way, and the piano was a leading item in the repertoire⑤.

As his career became increasing successful, Edison had developed a narrow range of philanthropic⑥ ideals, including the desire to place a piano in the home of every American. To ensure affordability, expensive, resonant⑦ wood would be replaced by a piano framework of concrete. Perfectly tuneable, if somewhat immovable, it was hoped the instrument would extend America's musical talent – although Edison was far from the best judge of this. He'd been pretty much deaf since childhood; the result, he claimed, of being picked up by the ears onto a moving train by a helpful conductor – presumably, as Edison's brother suffered similar partial deafness, a family

爱迪生的水泥家具

大概在处死大象的同一时期，**托马斯·阿尔瓦·爱迪生**就想出了一个扼杀音乐的计划。他的第一项发明是为美国国会议员设计的电报式投票机，但由于机器反应太慢，等开始工作时，议员们往往已经改变主意投给另一方了。这项发明失败之后，爱迪生决定从今往后只发明有明确需求的产品。在想了成百上千个主意之后，他想到了水泥家具，第一件产品就是水泥钢琴。

注释

托马斯·阿尔瓦·爱迪生（1847—1931），发明家、企业家。他发明的留声机、电影摄影机、电灯对世界有极大影响。他一生的发明共有2000多项，拥有专利1000多项。

随着事业不断取得成功，爱迪生产生了一丝狭隘的慈善理念，包括让每个美国家庭拥有一架钢琴的愿望。为了保证大家都能买得起，他用水泥钢琴代替了昂贵的共鸣木材质钢琴。他认为，这种钢琴虽然不易搬动，但音调完美，有望拓展美国人民的音乐才华——尽管他根本不具备评判音乐才华的能力，他从小耳朵就聋了。他说，这都是那个售票员干的好事，售票员抓着他的耳朵把他提到了开动的火车上，导致

hazard[①]. But Edison had a reason to develop concrete furniture that went beyond the philanthropic. As owner of the Portland Cement Company, he produced more concrete than he knew what to do with. The work to which the factory was being put when he formed it in 1899 – iron-ore production – wasn't working out and its heavy industrial equipment was set for scrap.

Here Edison's entrepreneurial[②] spirit came to the fore. Urbanisation was continuing apace[③] in the US, yet the standard of housing for the working classes was generally poor. Edison, who had grown up in poverty, dreamed of ridding[④] America of 'city slums' by building inexpensive concrete houses for workers. He aimed to construct a house that could sell for no more than $1,000. And not just concrete houses, but concrete houses equipped with concrete furniture (at a maximum cost, he calculated, of $200 to kit out the home). An altruistic[⑤] venture aimed at alleviating[⑥] the lot of the poor, Edison envisaged[⑦] workers relaxing in the evenings with a sing-song round their concrete piano. Concrete, he declared, 'will make it possible for the labouring man to put furniture in his home more artistic and more durable than is now to be found in the most palatial[⑧] residence in Paris or along the Rhine'.

Within four years, Edison's factory was turning out cement[⑨] on a massive scale: 3,000 barrels[⑩] a day by 1905. Highly mechanised, with enormous kilns[⑪], it was soon one of the biggest cement factories in the world. Combined with his ready eye for an invention, Edison realised that as concrete was durable and flexible, at least before it set, if he reinforced it with a mesh[⑫] of metal rods[⑬], he could make ever larger structures. Not only could he build cheap, good homes for workers, whole houses could also be moulded in one piece. This process, which cast an entire house from one mould[⑭], rather than the traditional

注释

① hazard ['hæzəd] *n.* 危险
② entrepreneurial [ˌɒntrəprə'nɜːrɪəl] *adj.* 具有创业素质的
③ apace [ə'peɪs] *adv.* 快速地
④ rid [rɪd] *v.* 使（某地、某人）摆脱
⑤ altruistic ['æltruːˌɪstɪk] *adj.* 利他的
⑥ alleviate [ə'liːvɪˌeɪt] *v.* 减轻（不适）
⑦ envisage [ɪn'vɪzɪdʒ] *v.* 设想
⑧ palatial [pə'leɪʃəl] *adj.* 宫殿般的
⑨ cement [sɪ'mɛnt] *n.* 水泥
⑩ barrel ['bærəl] *n.* 桶
⑪ kiln [kɪln] *n.* 窑
⑫ mesh [mɛʃ] *v.* 网眼织品
⑬ rod [rɒd] *n.* 长杆，长棒
⑭ mould [məʊld] *n.* 模子

了他的耳聋。但是爱迪生的哥哥也一样部分失聪，所以估计这是家族遗传病。不过除了出于慈善目的外，爱迪生还有一个更好的理由去发明水泥家具，那就是他的波特兰水泥公司生产了太多水泥，不知道该怎么处理。1899年建厂之初是为了生产铁矿石，但没成功，笨重的工业设备一直闲置在那里。

爱迪生的创业精神在此显现出来。美国城镇化迅速发展，而工薪阶层的住房水平通常很低。出身贫苦之家的爱迪生梦想着为工薪阶层建造价格低廉的水泥房子，使美国不再有“城市贫民窟”。他的目标是建造售价不超过1000美元的房子。不仅是水泥房子，还是配有水泥家具的水泥房子（他计算了一下，一栋房子的家具最多花费200美元）。爱迪生这次无私的冒险是想缓解贫困人群的压力，他设想着工人们晚上下班后可以围着水泥钢琴唱歌放松。他说：“水泥可能使劳动人民在家里放置的家具比现在巴黎或莱茵河畔豪宅里放置的家具更具有美感，更结实耐用。”

四年时间里，爱迪生的工厂生产了大量的水泥，到1905年，每天生产3000桶水泥。因为高度机械化，加上巨大的窑炉，它很快就成了世界上最大的水泥厂之一。在那双随时盯着发明的眼睛的帮助下，爱迪生意识到水泥结实耐用，而且在浇铸前容易塑形，如果用钢筋网加固，就能建造出更大的建筑。他不仅可以为工人建造物美价廉的房子，而且还可以整体浇铸房子。这种工艺采用一个模型浇铸整栋房子，不同于传统的分别垒墙然后再连接到一起的方法，造价虽然较高，但同时打开了另一个市场——昂

注释

① bolt [bəʊlt] *v.* 用插销闩上，能被闩上，用螺栓把（甲和乙）固定在一起
② impoverish [ɪm'pɒvərɪʃ] *v.* 使贫困
③ aggregate ['ægrɪgɪt] *adj.* 合计的
④ hydraulic [haɪ'drɒlɪk] *adj.* 液压的
⑤ phonograph ['fəʊnəˌgrɑːf] *n.* 留声机
⑥ cabinet ['kæbɪnɪt] *n.* 储藏橱，陈列柜

way of constructing individual walls and bolting[①] them together, was costly, but opened up another market as well – expensive concrete homes. Ugly but pricey, the wealthy could proudly invest in his properties too. But whether for impoverished[②] workers or affluent executives, pouring concrete consistently and reliably isn't easy, particularly when the mould has to make individual rooms all in one go. The mixture of the aggregates[③] and water has to be just right, and hydraulic[④] force has to push the concrete vertically in parts. The delicate process didn't always run smoothly.

Nonetheless, commercial projects started to be commissioned: the world's first concrete highway, numerous buildings in New York City and the Yankee football stadium amongst them. The first of the philanthropic homes for workers were cast in Huxton Street in South Orange, New Jersey, in 1911. By 1917 just eleven concrete homes had been constructed, yet not one was sold, even at $1,200 (only slightly more than Edison's original aim); a bargain price for a home at the time. Writing in *Collier's Weekly* in the 1920s, architect Ernest Flagg said: 'Mr. Edison was not an architect. It was not cheapness that people wanted so much as relief from ugliness and Mr. Edison's early models entirely did not achieve that relief.' There were practical problems of living in a concrete home too: one can go through endless nails hanging pictures on a concrete wall and householders who wanted to adapt the accommodation discovered the impossibility of taking a wall out when the house is cast from a single mould.

With only some hundred concrete houses sold, but a factory to keep in action, Edison stepped up production of concrete furniture. At the 1911 annual convention of American Mechanical Engineers, he announced a plan for bedroom sets retailing at $5; baths, phonograph[⑤] cabinets[⑥] (record players)

贵的水泥房屋。样子难看但很昂贵，富人也会自豪地买下这些房屋。然而，无论是对于穷困的工人还是对于富有的高级管理人员来说，浇注水泥并不容易，尤其是模型要求所有房间都一次成型。水泥骨料和水的混合比例要恰到好处，水压要分几部分向水泥垂直施压。这个工艺总是无法顺利完成。

▲发明家托马斯·爱迪生和他的“革命性”的水泥房屋模型。他相信这一创新将提供大量的低价房屋，他甚至还设计了与之配套的水泥家具。

Inventor Thomas Edison with a model of his 'revolutionary' concrete house. He believed the innovation would provide plentiful low-cost housing and even designed concrete furniture to go inside the homes.

尽管如此，商业计划仍然开始启动，包括世界上第一条水泥高速公路、纽约市的大量建筑、扬基橄榄球场。第一批为工人修建的慈善房屋于1911年在新泽西州南橘子郡的霍克斯顿大街落成。到1917年，只建成了11座房屋，但一座也没有卖出去，哪怕售价只有1200美元（这个价格只比爱迪生最初预期高了一点儿），在当时来说，已经很便宜了。建筑师欧内斯特·弗拉格（Ernest Flagg）在20世纪20年代出版的《科里尔周刊》中写道：“爱迪生先生不是建筑师。人们不会因为价格低就完全不在意外观，而爱迪生先生早期的模型完全无法达到人们对外观的要求。”在水泥房子里居住还有一些现实问题：房主在墙上挂画时可能要没完没了地砸钉子，而想要改造空

trimmed[①] in white and gold, and a piano. Just like his elephant stunt, everything was done with an eye to press coverage. For one press wheeze[②], concrete phonograph cabinets were shipped with a sign saying: 'Please drop and abuse this package'. He told the *New York Times* that the phonograph cabinet provided better acoustic qualities than wood. The concrete piano housed a wooden soundboard[③] – there was no getting away from that – and to Edison's partially deaf ear, the tone would be indistinguishable from the traditional instrument.

After all, unlike wood, which rots[④] and splinters[⑤], concrete, he claimed, was perfect for furniture. Edison did concede[⑥], however, that concrete reinforced with steel rods would probably make for an uncomfortable sofa. So he set to work amending[⑦] the constituency[⑧] of his concrete, suffusing[⑨] it with air to make 'foam concrete'; an oxymoron[⑩] that fooled no one. The price – though less than half of that of equivalent furniture – was surely attractive. It would also be an investment; newlyweds[⑪] would have furniture that would last longer than their marriage. It may be heavy, but Edison felt that was part of its attraction, and

注释

① trim [trɪm] *v.* 镶边于
② wheeze [wiːz] *n.* 喘息声，呼哧呼哧声
③ soundboard ['saʊn(d)bɔːd] *n.* 响板，共鸣板
④ rot [rɒt] *v.* 腐烂
⑤ splinter ['splɪntə] *v.* （使）裂成碎片
⑥ concede [kən'siːd] *v.* （常指不情愿地）承认
⑦ amend [ə'mɛnd] *v.* 修正
⑧ constituency [kən'stɪtjʊənsɪ] *n.* 一批顾客（或支持者，拥护者，赞助者）
⑨ suffuse [sə'fjuːz] *v.* 遍布
⑩ oxymoron [ˌɒksɪ'mɔːrɒn] *n.* 矛盾修饰法
⑪ newlywed ['njuːlɪˌwɛd] *n.* 新婚夫妇

间的房主会发现房子是整体浇铸的，根本没法实施。

虽然只售出了几百套水泥房子，但水泥厂还在正常运转，爱迪生又开始生产水泥家具。在1911年的美国机械工程师年会上，他宣布计划制造卧室的成套家具，零售价仅为5美元，包括浴缸、白色和金色镶边留声机（电唱机）与钢琴。和上次的大象噱头一样，他所做的一切只是为了得到媒体的关注。一家报纸开玩笑说，水泥留声机在发货时贴着这样的标签："请随意装卸本包装箱"。爱迪生告诉《纽约时报》，这台留声机比木质留声机的音效还好。水泥钢琴里装了木质共鸣板——没有摒弃木材——对于部分失聪的爱迪生来说，这种钢琴的音色与传统钢琴没什么区别。

他说，水泥毕竟不像木材那样容易腐烂和破裂，是做家具的完美材料。然而，他也承认水泥经过钢筋加固，可能使水泥沙发不舒服。于是为了弥补顾客，他开始努力改进水泥家具，在里面填充空气，制成"泡沫水泥"，但这种矛盾的说辞，愚弄不了任何人。然而水泥家具的价格还不到同等家具的一半，这点还是很吸引人的。这也可作为一笔投资，新婚夫妇将会买到比他们婚姻还要长久的家具。这种家具可能十分笨重，但爱迪生认为这也是它吸引人的原因之一，有了新的泡沫水泥，它只比同等的木家具重25%。经过细心打磨，涂上特殊油漆后，外观也好看了。水泥家具的外观可以达到顾客要求的任何一种木家具的水平。

到84岁去世之时，爱迪生已经有了1093个专利。

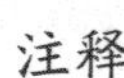

注释

① spit [spɪt] *n.* 口水

② incandescent [ˌɪnkæn'dɛsənt] *adj.* 遇热发光的，白炽的

③ bulb [bʌlb] *n.* 电灯泡

④ stencil ['stɛnsəl] *n.*（镂有图案或文字的）模版，蜡纸

with the new foam concrete it would only be about 25 per cent heavier than wooden alternatives. It would look good too. With a bit of spit[1] and polish and a special paint, concrete furniture could appear like any type of wood the purchaser required.

By the time he died at the age of 84, Edison had amassed 1,093 patents. He had invented or further developed products such as the incandescent[2] light bulb[3]; the phonograph; telegraph; a 'stencil[4]-pen' (the forerunner of a tattoo drill); the sewing machine and the X-ray. Other intriguing ideas, such as a device for communicating with the dead, never got beyond the scope of his imagination. But Edison's concrete furniture, his concrete piano and his concrete houses are not amongst his most notable successes. In Huxton Street, some of his concrete houses still stand today: concrete, after all, being nothing if not enduring.

他发明或改进的产品包括白炽光灯泡、留声机、电报机、“蜡纸铁笔”（文身针的前身）、缝纫机和X射线。他还有些有趣的想法，例如与死人交流的设备，不过只是停留在他的想象中了。然而，爱迪生的水泥家具、水泥钢琴和水泥房屋并不在他著名的成功发明之列。在霍克斯顿大街上，他建造的部分水泥房子至今还屹立在那里：毕竟水泥最大的优点也就是经久耐用了。

ABANDONED

THE X-RAY SHOE-FITTING MACHINE

注释

① fluoroscopy [flʊə'rɒskəpɪ] *n.* 荧光透视，荧光镜检查
② lethal ['liːθəl] *adj.* 致命的
③ fascinated ['fæsɪneɪtɪd] *adj.* 入迷的
④ sceptical ['skɛptɪkəl] *adj.* 表示怀疑的
⑤ fluorescent [ˌflʊə'rɛsənt] *adj.* 荧光的 *n.* 荧光
⑥ showmanship ['ʃəʊmənʃɪp] *n.* 夸张搞笑的主持才能

When German physicist Wilhelm Röntgen's 1895 discovery of X-rays (then known as Röntgen rays) won the 1901 Nobel Physics Prize,Thomas Edison, the inventor of the concrete house and a man always eager to build on scientific advancements, paid careful attention. Here was a concept worth developing. Within six months, the Edison X-ray tube was ready for market, and fluoroscopy① – the use of X-rays – began entertaining crowds as they gaped inside their own bodies and at the bones of hands, feet and head. Röntgen's X-rays and Edison's fluoroscope paved the way for an invention that was to change the way people were fitted for shoes, with lethal② effects for shop assistants, and not much good for the very feet they were designed to help either.

At the start, and indeed for decades following, the public greeted X-rays with awe. To be able to see into the body fascinated③ even the most sceptical④, as one man who complained that the fluorescent⑤ screen was nothing but a sheet of ground glass discovered to his embarrassment after being hauled onto the stage at its 1896 launch. Edison, deploying his customary showmanship⑥, forced the man to hold up his

X射线试鞋机

1895年，德国物理学家威尔海姆·伦琴（Wilhelm Rontgen）发现了X射线（当时被称为伦琴射线），因此而获得1901年的诺贝尔物理学奖。就在这时，水泥房子的发明者托马斯•爱迪生这位渴望在科学进步的基础上有所建树的人，开始对这个发现特别关注。这绝对是一个值得研究的想法！在短短6个月之内，爱迪生的X射线管就已准备投放市场。人们惊奇地发现，通过使用X射线的荧光镜，可以看到自己的身体内部以及手、脚和头部的骨骼。事实上，伦琴的X射线和爱迪生的荧光镜为一项发明铺平了道路，它将改变人们选鞋的方式，给店员带来生命危险，而且对那些试鞋的脚也没有多少好处。

起初，确切地说是一直到之后的几十年之内，公众始终怀着敬畏的心情接受着X射线。它能让你看见身体内部，连最挑剔的人也为之着迷。1896年，当荧光镜投放市场并向公众展示时，一名观众批评荧光镜屏幕只是块毛玻璃板。爱迪生用他惯有的表演天赋，让这个人举起手腕，

wrist[①] until he saw a hole through it, converting him instantly to the wonders of fluoroscopy. At the time, with no reason for concerns about the affect of the machines on health, everyone wanted a shot of their insides – one happy patient benefiting from a medical X-ray on the operating table on the very day of Edison's public exhibition. Soon the fluoroscope was enjoying commercial success in many forms; by the turn of the century being used to investigate complaints as complex as impotency and insanity[②].

Very quickly, however, fluoroscopy's darker side began to emerge. In 1900, Clarence Dally, Edison's colleague who had worked most closely on the fluoroscope, began to suffer skin lesions on the very area of his arm that he had habitually stuck into a machine during experiments. When skin grafts failed, first a hand was amputated and eventually both arms. Even this was insufficient to save him. Before long, cancer had taken him. A shocked Edison, who had almost lost his own eyesight, stopped working with fluoroscopes and began to warn of the dangers. 'Don't talk to me about X-rays,' he said. 'I am afraid of them … and I don't want to monkey with them.'

The concerns of an eminent[③], if eccentric[④], inventor, did little to prevent the fluoroscope's adaptation in ingenious ways though. Some products were undoubtedly useful: the work of Röntgen, Edison and later Marie and Pierre Curie created the basis of the science of radiotherapy[⑤] that remains in use today. Other ideas were little more than gimmicks[⑥], trifling[⑦] new products or marketing wheezes[⑧]. In 1924, in Milwaukee, USA, resisting protests from the Radiological Society of North America ('it lowers the profession of radiology'), surgical supplies salesman Clarence Karrer marketed a shoe-fitting fluoroscope. But failing to progress a patent application, Boston doctor Jacob Lowe beat him to the legitimate[⑨] claim of the

注释

① wrist [rɪst] *n.* 手腕
② insanity [ɪn'sænɪtɪ] *n.* 精神错乱
③ eminent ['ɛmɪnənt] *adj.* 卓越的，有名望的
④ eccentric [ɪk'sɛntrɪk] *adj.* 古怪的，异乎寻常的
⑤ radiotherapy [ˌreɪdɪəʊ'θɛrəpɪ] *n.* 放射治疗
⑥ gimmick ['gɪmɪk] *n.* 花招
⑦ trifling ['traɪflɪŋ] *adj.* 微不足道的
⑧ wheeze [wiːz] *n.* 喘息
⑨ legitimate [lɪ'dʒɪtɪmɪt] *adj.* 合法的

让他透过荧光镜屏幕看到自己的手腕上有一个洞。尴尬之余，这位观众对荧光镜下的奇迹感到万分惊奇。那时，人们还不知道这台机器会给他们的健康带来什么不良影响，每个人都想用它照一下自己的身体。就在爱迪生展示荧光镜那天，一位病人在手术台上荣幸地尝到了医用X射线的好处。不久，荧光镜就以各种形式获得了商业成功，到世纪之交甚至被用于诊断像性无能和精神错乱那样复杂的病情。

▲ 一台1927年发明的X射线试鞋机，曾用于华盛顿特区的博伊斯路易斯店，现陈列在该市的国家卫生与医学博物馆。

An X-ray shoe-fitting machine from 1927, used in the Boyce & Lewis store in Washington D.C., is now on display at the National Museum of Health and Medicine in the city.

然而好景不长，荧光镜的负面影响开始出现了。1900年，爱迪生的同事克拉伦斯·达利（Clarence Dally）的胳膊得了皮肤病，他与荧光镜的接触最频繁，发病部位正好就是他做实验时经常伸进机器的地方。植皮手术失败后，先是截去了他的一只手，后来又把两条胳膊也截掉了，但还是未能挽救他的性命。不久之后，癌症夺去了他的生命。爱迪生对此大为震惊，而他自己也几乎双目失明了。他停止了对荧光镜的研究，开始向周围的人警告荧光镜的危险。“不要跟我谈X射线，”他说，“我害怕它……我不想再摆弄它了。”

尽管如此，一位杰出但也许有些古怪的发明家对X射线的担忧，并没能阻止人们对荧光镜各种巧妙的改造，一些发明无疑十分有用。伦琴、爱迪生以及后来的居里夫妇的研究为今天仍在使用的放射疗法奠定了基础，另一些发明则仅仅是一些吸引人的噱头、无关紧要的新产品或营销策略。1924年，在美国密尔沃

注释

① apparatus [ˌæpəˈreɪtəs] *n.* 仪器，器械
② deform [dɪˈfɔːm] *v.* 使成畸形
③ bag [bæg] *v.* 占有，抢占
④ ambiguous [æmˈbɪgjʊəs] *adj.* 模棱两可的
⑤ snugly [ˈsnʌgli] *adv.* 舒适地，贴身地
⑥ footwear [ˈfʊtˌwɛə] *n.* 鞋类
⑦ viewfinder [ˈvjuːˌfaɪndə] *n.* 取景器

invention. Lowe, who had been working on a device since returning soldiers from the Great War turned up at his surgery with foot injuries, developed a machine that could X-ray feet without the need for boots to be removed. He applied for a patent in 1919, emphasising the medical advantage:

> With this apparatus[1] in his shop, a shoe merchant can positively assure his customers that they need never wear ill-fitting boots and shoes; that parents can visually assure themselves as to whether they are buying shoes for their boys and girls which will not injure and deform[2] the sensitive bone joints.

Quite coincidentally, by the time Lowe's US patent was granted, in the UK, the Pedescope Company of St Albans had bagged[3] their own patent. It had been 'in continuous daily use throughout the British empire for five years', they declared. So the claims to the first use of this ultimately dangerous invention are ambiguous[4].

By the end of the decade, the Pedescope Company and, in the US, the Adrian X-Ray Company of Milwaukee, were selling the devices widely. Almost every enterprising shoe shop had one: a large wooden box, nicely polished and presented, into which customers would place one foot at a time to see how snugly[5] new footwear[6] fitted. The length of the exposure varied depending on the size of the foot – the highest intensity for men, the middle one for women and the lowest for children. A timer set exposure time from five to forty-five seconds, the most common dose being twenty seconds. Shop assistant, child and parent would gape through one of three viewfinders[7] at the bones of the feet and the outline of the shoe, and a quick wiggle of the toes would reveal growing-room. The only shield between

基，手术器械推销员克拉伦斯·卡勒（Clarence Karrer）不顾北美放射学会的反对（他们的反对意见是："它贬损了放射学的尊严"），开始推销一种检查鞋子是否合脚的荧光镜。但是他没有成功申请到这个发明的专利，波士顿医生雅各布·罗维（Jacob Lowe）抢先一步成了这项发明的合法所有者。自从他的手术室出现从世界大战战场回来带着脚伤的士兵，罗维就开始研制一种无须脱鞋就能用X射线透视脚的机器。1919年他申请专利时，强调了这项发明对人体健康的好处：

> 如果鞋店里有了这种仪器，卖鞋的人就能肯定地向顾客保证，他们绝不会穿上不合脚的鞋靴；父母也能亲眼确保他们给子女买的鞋不会伤脚，使敏感的骨关节变形。

相当凑巧的是，罗维申请到这项美国专利时，圣奥尔本斯的足部光学仪器公司也在英国获得了专利。这家公司声称，他们的这项发明"在大英帝国已经普及使用五年了"。因此究竟是谁第一个使用这种事实上非常危险的发明，谁也说不清楚。

到20世纪20年代末，足部光学仪器公司和美国密尔沃基阿得雷恩X射线公司已经在市场上大量销售这种仪器。几乎每家想发达的鞋店都会有这样一台机器：一个擦得很亮、摆放得很好的巨大木箱，顾客可以一次放一只脚在里面，看新鞋是否合脚舒适。照射X射线的时间取决于脚的大小——男人照射的强度最大，女人居中，儿童最低。照

feet and tube was a miniscule[①] aluminium[②] filter. Shoe shops with pedescopes had a marketing advantage. Reluctant children now looked forward to trying on shoes, and it was the sign of a responsible parent. Remember, yelled a press advertisement for the Adrian X-Ray machine: 'They'll need their feet all through life.'

Buoyed[③] by success and the sound of ringing tills[④], marketing intensified. The Adrian X-Ray Company implored stores to 'place the machine in the most desirable location, [facing] the ladies' and children's departments by virtue of the heavier sales'. Press advertisements assured parents that: 'shoes that fit well, last longer.' And a well-shod[⑤] foot was better for health; safer: 'The Adrian special shoe fitting machine has been awarded the famous Parent's Magazine Seal of Commendation … a symbol of safety and quality to millions of parents all over America.'

Any concerns about dangers were slow to develop. But towards the end of the 1940s, with the machines now deployed for more than twenty years, reports of skin and bone marrow damage and growth problems were emerging. After two atomic bombs brought an end to the Second World War, the effects of high radiation dosage[⑥] were becoming recognised. In 1950, researchers Leon Lewis and Paul E. Caplan asked whether the shoe-fitting fluoroscope was hazardous[⑦]. 'Physicians and health physicists everywhere are beginning to be concerned about every potential source of radiation … quite irrespective[⑧] of the potentialities of the atomic bomb.' But even at this time, they said: 'The shoe-fitting fluoroscope is not an instrument with obviously hazardous potentialities. It has long been used and no direct clinical evidence of harm has yet been established.' So the machines continued to sell. By the middle of the century, more than 10,000 were in use in the US and a slightly smaller number

注释

① miniscule ['mɪnɪskjuːl] *adj.* 极小的（同 minuscule）
② aluminium [ˌæljʊ'mɪnɪəm] *n.* 铝
③ buoy [bɔɪ] *v.* 使振奋
④ till [tɪl] *n.*（商店等的）收银台，收银机
⑤ shod [ʃɒd] *adj.* 穿着……鞋的（shoe的过去式及过去分词）
⑥ dosage ['dəʊsɪdʒ] *n.* 剂量
⑦ hazardous ['hæzədəs] *adj.* 有危害的
⑧ irrespective [ˌɪrɪ'spɛktɪv] *adj.* 不考虑的

射时间通过计时器设置，范围从5秒到45秒，常用的是20秒。店员、儿童和家长分别通过三个取景器观看脚骨和鞋的外形，快速扭动一下脚趾还可以看到鞋内多余的空间。脚与射线管之间的唯一屏障仅是一块很小的铝制滤光板。有了这种X射线试鞋机，鞋店就能更好地推销产品。

▲ 荧光镜上供店员、家长和孩子观看的取景器。

The fluroscope had viewfinders for the shop assistant, parent and child.

现在，不愿试鞋的孩子也盼望着去鞋店试鞋；能带孩子去鞋店买鞋，家长也显得更有责任心。一则宣传阿得雷恩X射线机的广告这样喊道："记住，孩子一辈子都离不开他们的脚。"

X射线试鞋机被成功地推向市场，在利益的驱使下，制造商加紧推销这种机器。阿得雷恩X射线公司请求每个商店"把试鞋机放在商店里的最佳位置，即（面向）女鞋部和童鞋部，因为这两个部门的销量较大"。X射线试鞋机的广告宣传则向父母强调："合脚的鞋，穿得更久。"此外，穿合脚的鞋也更有利于健康、更安全，正如这句广告词所言："阿得雷恩特有的X射线试鞋机成为著名的《父母》杂志的推荐品牌……全美亿万父母认可的安全和质量的象征。"

虽然对于危险的担忧来得并没有那么快。但是到20世纪40年代末，这种装置在市场上已经存在了20多年的时候，有关皮肤和骨髓病变以及生长问题的报道便开始出现了。在两颗原子弹结束了第二次世界大战后，人们也逐渐认识到高辐射剂量所带来的后果。1950年，研究者莱昂 · 刘易斯（Leon Lewis）和保

注释

① detrimental [ˌdɛtrɪ'mɛntəl] *adj.* 有害的
② pinch [pɪntʃ] *v.* 掐，拧
③ lesion ['liːʒən] *n.* 损伤，伤口
④ dermatitis [ˌdɜːmə'taɪtɪs] *n.* 皮炎
⑤ amputate ['æmpjʊˌteɪt] *v.* 截（肢）
⑥ precaution [prɪ'kɔːʃən] *n.* 防范行动
⑦ backfire [ˌbæk'faɪə] *v.* 事与愿违
⑧ lag [læg] *v.* 落后

in the UK, Canada and Europe.

Medical evidence about the detrimental[①] effects did emerge. More than one shop assistant, used to putting their hands under the X-ray to pinch[②] around the shoe to confirm the fitting, suffered hand lesions[③] or dermatitis[④]. Injury to reproductive organs from prolonged exposure was also reported; one shoe model had to have a leg amputated[⑤]. Children were at greatest risk. Throughout the 1950s and '60s, US regulators gradually restricted the advisory amount of radiation to which customers, and children in particular, could be legally exposed. Only twelve doses per child per year, several states recommended – which still allowed a lot of shoe fitting. Some precautions[⑥], however, backfired[⑦]. When the American Conference of Governmental Industrial Hygienists (ACGIH) suggested a uniform set of standards in 1950, manufacturers met the regulations and then boasted about it in a way that suggested that pedescopes were government approved. Nonetheless, by the 1960s, thirty-three states had banned the device altogether, while in others only a licensed physician could operate them. Most shoe shops, being reluctant to hire a doctor to fit shoes (and face parents wondering why a medic was suddenly required to do something that had previously undertaken by a teenage boy earning pocket money), removed the machines.

The UK lagged[⑧] behind America. While US states fell over themselves to protect customers, British regulators resisted. As late as 1960, the Pedescope Company of St Albans was defending their machine against an attack in New Scientist magazine: 'The only real way of getting a higher standard of shoe fitting is to let both the customer and the sales person see inside the shoe the damage that is being done by a badly fitting shoe,' wrote its representative. 'The difficulty of shoe-fitting being a "blind" operation whose effect whether good or bad

罗·E. 卡普兰（Paul E. Caplan）对这种用于试鞋的荧光镜的安全性提出了质疑："各地的医生和健康物理学家开始关注各种可能的辐射源……它们全然不顾原子弹的潜在后果。"即便此时，他俩还是说："这种荧光镜没有明显的潜在危险。它已被使用很长时间，临床上没有直接证据表明它是有害的。"就这样，X射线试鞋机继续在市场上销售。到20世纪中期，美国有1万多台正在使用的试鞋机，英国、加拿大和欧洲则略少一些。

关于这种X射线机危害的医学证据确实出现了。不止一个店员的手出现病变或皮炎，这些店员常常将手放到X射线下按捏鞋子，看鞋子是否合适。长时间照射X射线给生殖器官带来损害的报道也时有出现，一位鞋模的一条腿还被截了肢。风险最大的还是孩子们。20世纪50—60年代，美国的监管者逐渐限制了对顾客特别是对儿童的法定建议辐射量。一些州建议儿童每年的最大辐射量为12个剂量——在这个标准下许多试鞋机仍然可以使用。然而，一些措施起了反作用。1950年，在美国政府工业卫生学家会议（ACGIH）制定了一套统一标准后，厂家对产品做出相应调整，向市场吹嘘自己的产品是经过政府批准生产的。不过，到了20世纪60年代，美国33个州已经完全禁止使用这种装置，在其他州只有持照医生才能操作这种装置。大部分鞋店由于不愿意雇医生给顾客试鞋（家长一定会问，为什么原来十来岁挣零花钱的男孩做的事，现在突然让医生来做），就不再使用这种机器了。

英国比美国反应慢了一些。当美国政府急匆匆地保护消费者的时候，英国的监管者却采取了抵制态度。20世

注释

① infallible [ɪn'fæləbəl] *adj.* 从不出错的

② radium ['reɪdɪəm] *n.* 镭

only becomes apparent later on is completely overcome by the use of a pedescope.'

Although some machines were still in use in the 1970s – and one was in service in West Virginia as late as 1981 – the tide was turning. What fifty years earlier had appeared to be exciting, modern and infallible[①] was, by then, a potential source of harm, even death. But throughout that period, even more remarkable radioactive inventions were on the market, including nuclear foods and drinks. Their origins were in a form of radioactivity called radium[②]. And it is that to which we turn next.

纪60年代末，圣奥尔本斯的足部光学仪器公司在《新科学家》杂志上为他们的机器辩护道："看鞋是否合脚的唯一真正的更高标准是，让顾客和售货员看看鞋里面，看看不合脚的鞋是怎样对脚造成伤害的。"文章写道："试鞋是'盲目的'，新鞋要穿上一段时间才知道是好是坏，而试鞋机完全可以解决这个问题。"

虽然20世纪70年代仍有一些X射线试鞋机被使用，而直到1981年西弗吉尼亚州还有一台工作着，但试鞋机大势已去。50年前看上去新奇、现代、绝对可靠的机器，这时人们已经知道它会带来危害，甚至死亡。不过，在试鞋机盛行的那段时间，市场上还有更引人注目的含放射性物质的发明，包括核子食物和饮料，它们起源于一种叫镭的放射性物质。我们下一个故事会讲到它。

BANNED
THE CURE THAT KILLED

注释

① emit [ɪ'mɪt] *v.* 发出，散发（热、光、气体或气味）
② unwittingly [ʌn'wɪtɪŋli] *adv.* 不知情地，无意识地
③ craze [kreɪz] *n.* 一时的狂热
④ radioactive [ˌreɪdɪəʊ'æktɪv] *adj.* 放射性的
⑤ gadget ['gædʒɪt] *n.* 小器械（有时暗指复杂、不必要的东西）
⑥ luminous ['luːmɪnəs] *adj.* 发亮的
⑦ intriguing [ɪn'triːgɪŋ] *adj.* 新奇的
⑧ silhouette [ˌsɪluː'ɛt] *n.*（强光或浅色背景衬托下的）黑色轮廓
⑨ emanate ['ɛməˌneɪt] *v.* 发散出（品质）

In the last decade of the nineteenth century, new discoveries in physics came thick and fast: Heinrich Hertz's high frequency radio waves in 1886; Wilhelm Röntgen's shortwave equivalent, X-rays, in 1895; and Henri Becquerel's identification of spontaneous radiation shortly afterwards. Then, in 1898, Marie and Pierre Curie discovered a new substance that emitted① radiation. Coining the term 'radioactivity' they named the element radium – and unwittingly② started a craze③ for radioactive④ gadgets⑤, beverages and quack cures.

Once discovered, the Curies quickly set to work to truly understand how radium was constituted. From a shed at the school at which Pierre had a workshop, they examined the element intensively; the luminous⑥ substance emitting an intriguing⑦ blue glow that warmed Marie's heart in more ways than she realised at the time. It was joyous, she said, to go down to their shed at night and see 'from all sides the feebly luminous silhouettes⑧' emanating⑨ within. The light – and the radiation – that radium emitted could be seen through the thickest substances. On the evening they jointly won the 1903 Nobel Physics Prize with Henri Becquerel, Pierre entertained

致命的疗法

19世纪的最后10年，物理学领域的新发现层出不穷：1886年，海因里希·赫兹（Heinrich Hertz）发现了高频无线电波。1895年，威廉·伦琴（Wilhelm Röntgen）发现了一种短波电磁波——X射线。不久，亨利·贝克勒耳（Henri Becquerel）发现了自发辐射。1898年，居里夫妇（Marie Curie, Pierre Curie）发现了一种能发出放射线的新物质，他们把这种元素命名为镭（Radium），并创造了术语“放射性”，无意中让放射性的小玩意儿、饮料和骗人的疗法风靡一时。

发现镭以后，居里夫妇立即着手研究镭的结构。皮埃尔在学校的一间棚屋内有一个工作室，他们在这里对这种元素开始了紧锣密鼓的研究。这种发光的物质能发出一种神秘的蓝光，这让玛丽的心暖洋洋的，尽管她当时没想到镭能给她带来什么。她说每当晚上去棚屋，看到四面都是发出微光的各种实验器皿的轮廓时，她就很开心。事实上，镭放射出来的光和放射线能穿透最厚的物质而被看

注释

① waistcoat ['weɪsˌkəʊt] n.（西装）背心，马甲
② throb [θrɒb] v. 阵痛
③ alight [ə'laɪt] adj. 燃烧着的
④ perennial [pə'rɛnɪəl] adj. 不断出现的，长期存在的（问题、困难）
⑤ carnotite ['kɑːnəˌtaɪt] n. 钾钒油矿
⑥ camelhair ['kæm ə lheə] n. 驼绒，骆驼呢
⑦ repetitive [rɪ'pɛtɪtɪv] adj. 重复（因而乏味的）
⑧ mischievous ['mɪstʃɪvəs] adj. 调皮的
⑨ slap [slæp] v. 掴，（用手掌）拍打
⑩ witless ['wɪtlɪs] adj. 傻的，愚蠢的
⑪ orthodontic [ˌɔrθə'dɑntɪks] adj. 畸齿矫正的，齿列矫正的
⑫ luminous ['luːmɪnəs] adj. 发亮的
⑬ dial ['daɪəl] n. 刻度盘
⑭ decay [dɪ'keɪ] v. 腐坏
⑮ diagnose ['daɪəgˌnəʊz] v. 诊断
⑯ grotesque [grəʊ'tɛsk] adj. 荒唐的，怪异的
⑰ palate ['pælɪt] n. 上颚
⑱ erode [ɪ'rəʊd] v. 侵蚀

their friends by producing a little tube from his waistcoat[①] which, throbbing[②] with a blue light, illuminated the small party of beaming faces. He wasn't aware of the implications – and, being run over by a horse-drawn carriage three years later, never lived to see it kill Marie – but the Curies had invented glow-in-the-dark products and they were to set the world alight[③].

The perennial[④] puzzle of how to tell the time in the middle of the night was among those solved by the Luminous Materials Corporation, which from 1915 to 1926 produced radium from a mineral called carnotite[⑤]. Young girls in their first jobs out of school mixed radium powder with glue and water and, with fine strokes from a camelhair[⑥] brush, painted luminous hands on watches and clocks. When the brushes lost their fineness, which they did very quickly, the girls pointed them back into shape with their lips. The taste of radioactive paint aside, the work, while repetitive[⑦], could be fun. When one of the teenagers discovered that her handkerchief glowed in the dark after she'd blown her nose, it opened all kinds of possibilities for mischievous[⑧] entertainment. By slapping[⑨] the mixture on their teeth and faces, boyfriends could be frightened witless[⑩] when the lights went out.

At least, to begin with, the girls had teeth that could be painted. This happy orthodontic[⑪] pleasure wasn't to last long, as factory worker Grace Fryer was to discover. After spending three years painting luminous[⑫] dials[⑬] at the New Jersey factory, Grace's teeth were in trouble. Within two years of leaving, her health was in serious decline and, although she didn't yet know it, she was dying. Loss of her teeth was just the start of her problems. Soon bone decay[⑭] in her mouth and back was diagnosed[⑮]. Grace wasn't alone in such misfortune. Most of her seventy colleagues were by now also suffering grotesque[⑯] illnesses. One woman's palate[⑰] 'had eroded[⑱] so that it opened

见。在居里夫妇与亨利·贝克勒尔（Henri Becquerel）共同获得1903年诺贝尔物理学奖的那天晚上，皮埃尔从他的马甲里掏出一个小管子给他的朋友们欣赏。管子里跳动的蓝光照亮了每个人的笑脸。虽然皮埃尔不知道这个发现意味着什么——（三年后他被马车轧死，没能活着看到玛丽是怎样被这种物质致死的）但是他知道，他们发现的这种能在黑暗中发光的东西轰动了全世界。

如何在半夜看清时间是一个长久困扰人们的问题，这是发光材料公司所解决的众多问题中的一个。1915—1926年，这家公司从一种称为钒钾铀矿的矿石中提取镭。刚出校门的年轻女工将胶、水和镭粉混合起来，然后用驼绒刷子在钟表上勾勒出发光的指针。很快，刷毛就变得不整齐了，女工就会用嘴将刷毛理齐。除了这种含放射性物质的涂料味道不太好，这个工作还蛮有趣的，虽然有点重复。当一个十几岁的女工发现她的手帕在黑暗中发亮时，她们开始玩儿各种恶作剧游戏。她们把涂料拍在自己的牙齿和脸上，关上灯，吓唬自己的男朋友。

至少，一开始这些女孩还可以在自己的牙齿上涂颜料。然而，格蕾丝·弗莱尔（Grace Fryer）发现，这种有趣的牙齿“美容”游戏玩儿不了多久。弗莱尔是新泽西工厂的一名工人，她在那里涂了三年的夜光表盘，牙齿就出了问题。在两年的休假期间，她的健康状况日益恶化，尽管她还不知道自己即将死去。牙齿脱落仅仅是一个开始。很快她又被诊断出嘴部和背部的骨骼出现腐烂。格蕾丝不是唯一遭受这种不幸的人。此时，她的70位同事中的大部分人也正遭受各种怪异的疾病。据报道称，一位女性的上

注释

① nasal ['neɪzəl] *adj.* 与鼻子有关的
② radioactive [ˌreɪdɪəʊ'æktɪv] *adj.* 放射性的
③ ruthless ['ruːθlɪs] *adj.* 残酷的
④ woefully ['wəufəli] *adv.* 悲伤地，不幸地，使人痛苦地
⑤ exhume [ɛks'hjuːm] *v.* 挖掘（尸体以供检验）
⑥ inhale [ɪn'heɪl] *v.* 吸入，吸气
⑦ amorous ['æmərəs] *adj.* 性欲的
⑧ trial ['traɪəl] *n.* 审判
⑨ zing [zɪŋ] *n.* 活力，趣味
⑩ wondrous ['wʌndrəs] *adj.* 奇异的，令人惊叹的
⑪ hair tonic 护发素
⑫ condom ['kɒndɒm] *n.* 安全套
⑬ pharmaceutical [ˌfɑːmə'sjuːtɪkəl] *adj.* 制药的
⑭ antiseptic [ˌæntɪ'sɛptɪk] *n.* 消毒剂
⑮ malaria [mə'lɛərɪə] *n.* 疟疾
⑯ indigestion [ˌɪndɪ'dʒɛstʃən] *n.* 消化不良

into her nasal[①] passages', said a report. 'Tests showed her to be radioactive[②].'

Five women launched legal action against the company, now known as the US Radium Corporation. In the face of a ruthless[③] campaign of misinformation by a firm determined to resist each $250,000 claim, the judge found in favour of the workers, but awarded just $10,000 in immediate cash and a smaller annual amount for life. Those lives turned out to be woefully[④] short. By 1928, thirteen women were dead, needlessly young, and scores were to follow. One body, that of Amelia Maggia, suffered the indignity of being exhumed[⑤] for further medical evidence during the trial; her bones found to be 'still luminous with the radium she swallowed'. Investigations many years after the event showed that even the dust on the factory floor was radioactive and an experiment with cats revealed how their bones decayed from inhaling[⑥] it. Ninety years after the radium girls first painted watch dials, radiation levels were still being assessed at the site.

In many respects, the women were fortunate to win their case against a company manufacturing a product that, contrary from being a health risk, was seen as positively health enhancing. Well before the factory girls' glowing teeth were alarming amorous[⑦] boys, and even throughout the trial[⑧] and well beyond, companies were adding zing[⑨] to their products by including the wondrous[⑩] ingredient. Hair tonic[⑪], sweets, toothpaste, toys, condoms[⑫], bread: all now with added radium. But no sector embraced radium more than medicine. Almost as soon as the Curies announced their discovery, pharmaceutical[⑬] companies began to market radium ruthlessly. As an antiseptic[⑭] cream, a malaria[⑮] cure, the very thing you needed if you suffered from kidney or liver disease, indigestion[⑯], or even insanity; radium would help. All symptoms could be relieved, of course, by

颚“发生糜烂，缺口直通到鼻腔”“试验表明她本人具有放射性。”

5名女工起诉了这家公司——那时已更名为美国镭公司。她们指控这家公司没有给员工提供正确的信息，并向其索赔每人25万美元，公司坚决拒绝赔偿。这场官司打得非常激烈，结果工人胜诉，但每人只得到1万美元的现金赔偿和有生之年每年一小笔钱款。不幸的是，这些女工没活多久。到1928年，13位年轻的女工已经死去，后来又有几十个人死去。审判期间为了提供更多的医学证据，一位名叫阿米莉亚·马贾（Amelia Maggia）的女工的尸体在下葬后又不幸被重新挖出来，结果发现在她骨头中，“吃下去的镭仍在闪闪发光”。许多年以后又有人做了调查，发现连工厂地板上的灰尘都具有放射性。实验表明，猫吸入了这种灰尘后，骨骼出现腐烂。自这些镭女郎开始涂表盘90年后，人们仍一直在测量那里的辐射水平。

话又说回来，这些女工能打赢这场官司还算幸运，因为这家公司的产品不仅不被视为对健康有害，还被认为能给身体带来好处。在女工用发光的牙齿来吓唬亲密男友很早以前，甚至在官司期间和之后很长一段时间，一些公司为了给它们的产品增加活力，会在产品中加入这种神奇的成分。护发素、糖果、牙膏、玩具、避孕套、面包，这些里面都添加了镭。不过没有哪个领域会比医药领域更为广泛地使用镭。几乎就在居里夫妇宣布他们发现的同时，制药公司就开始拼命地在市场上推销镭。它可以被用来制作抗菌药膏，还可以治疗疟疾、肾病、肝病、消化不良，甚至精神错乱。当你洗完一次镭浴后，所有症状都能得到缓

taking a radium bath. Arthritis or rheumatism[①] sufferers were advised to hold their faces over a bowl of boiling water, add a dash[②] of radium and inhale. Old, ugly or vain people had radium to thank for its skin-rejuvenating[③] properties, which wiped years off them. And if all this sounded like quackery[④], any concerns, and there were few, were assuaged by the science: the correlation between radium inhalation and improvement in arthritis and rheumatism sufferers, for example, had been proven, erroneously[⑤] as it turns out, as early as 1906.

The most popular product of all was 'liquid sunshine' (radioactive glowing water), launched in 1904 by Dr George F. Kunz and Dr William J. Morton, professors of electro-therapeutics[⑥] at the New York postgraduate medical school. To illustrate a talk on how to make luminous drinks out of radium, the doctors had managed to acquire for the evening the world's biggest diamond, the Tiffany diamond. Placing a piece of radium behind the jewel, the room lit up. 'The light seen was like that of phosphorous[⑦],' said the New York Times. 'Afterwards the diamond was placed in a glass of water where it shone beautifully.' With 'liquid sunshine', the name the doctors admitted was chosen for its sales appeal, 'the whole interior of a patient could be lighted up'. Furthermore, the audience was told, radium may be the substance that gives the world's spring waters their curative powers. Even greater evidence was to follow. British scientist J.J. Thompson, who won the Nobel Physics Prize in 1906 for his discovery of the electron[⑧], had also found radioactivity in well water; the result of the presence of radium. Where radium (what today is known as radon) is present in the rocks over which water flows, the water was thought to be blessed with curative properties. A rush to the spas followed. Across America, health resorts[⑨] boasted of the radium they possessed naturally in their springs. In Europe, Marie

注释

① rheumatism ['ru:mə,tɪzəm] *n.* 风湿病
② dash [dæʃ] *n.* 少量，少许
③ rejuvenate [rɪ'dʒu:vɪ,neɪt] *v.* 使年轻
④ quackery ['kwækərɪ] *n.* 庸医的骗术
⑤ erroneously [i'rəuniəsli] *adv.* 错误地，不正确地
⑥ therapeutics [,θɛrə'pju:tɪks] *n.* 治疗学
⑦ phosphorous ['fɒsfərəs] *adj.* 磷的
⑧ electron [ɪ'lɛktrɒn] *n.* 电子
⑨ resort [rɪ'zɔ:t] *n.* 度假胜地，旅游胜地

解。他们还建议关节炎或风湿病患者把脸凑近一碗加入少许镭的开水，然后吸入蒸汽。镭还被认为具有活肤功能，年老、丑陋或是爱慕虚荣的人使用它可以抹去脸上岁月的痕迹。如果你觉得这些听起来像骗人的把戏，那么科学可以让你免除任何担忧（很少有人担忧）。例如，吸入镭能减缓关节炎和风湿病症状早在1906年就为科学实验所证实，尽管后来发现这是错误的认识。

在这些产品中，最流行的当属“液体阳光”（发光的放射性水）了，这是由纽约医学研究生院电疗法教授乔治·F. 昆茨（George F. Kunz）博士和威廉·J. 莫顿（William J. Morton）博士在1904年推出的。在一次晚间访谈节目中，为了展示如何用镭制作发光的饮料，两位博士找来了世界上最大的钻石——蒂芙尼钻石。当他们将一块镭放到钻石后面时，整个房间都被照亮了。根据《纽约时报》报道：“那种光看上去就像是磷发出的光。”后来这块钻石又被放入一杯水中，在水中发出美丽的光芒。饮下“液体阳光”（这是两位医生为提高销售吸引力而挑中的名字），“病人的整个身体内部都会被点亮”。接着他们又告诉观众，镭有可能就是使泉水具有治疗作用的物质。后来人们又找到了更多的相关证据。英国科学家J. J. 汤普森（J.J. Thompson，曾因发现电子而获1906年诺贝尔物理学奖）发现井水也具有放射性，原因就是里面有镭。如果水流过含镭（现在人们知道是氡）的石头，人们就认为水具有治病的作用。于是，人们又一窝蜂地去泡温泉。整个美国的疗养胜地都在吹嘘它们的温泉含有天然镭。在欧洲，居里夫妇也鼓励人们去泡温泉，以摄取能增强活力

注释

① spa [spɑː] *n.* 矿泉疗养地
② logistical [lə'dʒɪstɪkəl] *adj.* 在组织上的
③ crock [krɒk] *n.* 瓦罐
④ ceramic [sɪ'ræmɪk] *n.* 陶瓷制品
⑤ stiff [stɪf] *adj.* 硬的，不易弯曲的
⑥ tipple ['tɪpəl] *n.* 常喝的酒
⑦ magnate ['mægneɪt] *n.* 巨头，大亨

and Pierre Curie encouraged people to head to ematorias or inhalatorias, as spas[①] were known, to soak up refreshing radium and radioactivity.

But not everyone could get to, or afford, spa treatment. As radioactive water has to be drunk quickly if the active ingredient, radium, is not to lose its powers, bottling water raised logistical[②] issues. Instead, enterprising companies designed products that enabled customers to benefit from radium water at home. Amongst the first and most successful, in 1912 San Francisco company Revigator patented a 'radioactive water crock[③]' and, although its $29.95 price was out of reach for many people, sold hundreds of thousands. The 'crock', a ceramic[④] jug, contained radium which, when water was added and left overnight, would deliver 'the lost element of original freshness – radioactivity' to the water by breakfast. 'Fill jar every night. Drink freely. Average six or more glasses daily,' the packaging advised. It was 'a perpetual health spring in the home'. The effects of the refreshing radioactive beverage could be long lasting. In 2010, brave chemists at St Mary's University, Maryland, USA, bought Revigator jars on eBay and found them emitting as much radiation as they did almost a hundred years ago.

The Revigator faced stiff[⑤] competition from smaller, transportable, and, to the benefit of the radium-poor, much cheaper radium water products. Instead of water being placed in a jar, with the Thomas Cone, the Zimmer Emanator and the Radium Emanator, radium salts were dissolved in them. Americans especially were fond of such radioactive tipples[⑥], until stories began to emerge with alarming similarities to those of the luminous watch girls. Businessman Eben Byers, a millionaire steel magnate[⑦], knocked back 1,400 bottles of Bailey Radium Labs' Radithor in two years; his story

的镭和放射性。

不过，不是每个人都能去泡温泉，或者说能支付得起昂贵的温泉疗养费用的。而想要放射性水里面的活性成分镭发挥作用的话，就必须尽快饮用，因此瓶装水就出现了时效问题。于是，一些积极创新的公司设计出了能使消费者在家中就可以获得镭水的产品。最早最成功的一个产品是1912年旧金山公司瑞维给特申请的一项专利——放射性水罐。尽管它的价格高达29.95美元，但仍卖掉了几十万台。这种放射性水罐由陶瓷做成，里面装有镭，当加入水并存放过夜后，第二天早餐前水中就会含有“失去的原始新鲜元素——放射性”。产品包装上写着：每晚将罐子装满水。可随意饮用。通常每天饮用6杯以上。这个罐子是“家中永久的健康源泉”。当然，这种提神的放射性饮料效果非常持久。2010年，美国马里兰州圣玛丽大学一些勇敢的化学家在易趣网上购买了瑞维给特生产的这种水罐，发现它们与100年以前放射出的辐射几乎一样多。

此外，市场上还有许多体积较小、方便移动、价格便宜（这对穷人是个好消息）的镭水产品与瑞维给特水罐激烈竞争。像可以放入水中的托马斯锥、齐默尔放射器和镭放射器，它们中的镭盐可以直接溶解在水里，而不需要把水盛在一个罐子里。美国人特别喜欢这种放射性的提神饮料，直到与发光表女孩类似的事情开始出现。钢铁大亨埃本 · 拜尔斯（Eben Byers）在2年内喝下了1400瓶贝利镭实验室的雷地拓产品。他的故事最终登上了《华尔街日报》，标题是 “喝镭水喝到他的下巴都掉了”。直到那时，公众才开始关注这个问题。

culminating[①] in a Wall Street Journal article headlined: 'The Radium Water Worked Fine Until His Jaw Came Off '. Only then did the public begin to take notice.

Yet the notion of radium as a medicine was deeply ingrained. Throughout the 1930s, sales of radium water slowed as scare stories, and evidence, mounted[②]. Other new radium products, however, took their place: $150 would buy a 14-carat[③] gold Radiendocrinator male pouch[④], or condom[⑤], in its own velvet-lined[⑥] leather case. A nose cone respirator[⑦] was cheaper, but less enjoyable. For soldiers ready for battle in the Second World War, the glow-in-the-dark radium products never really went away: radium tacks placed in barbed[⑧] wire helped their comrades locate passage through, guns came with radium-lit sights for night-time aim, and in the trenches[⑨] troops wore luminous wristwatches[⑩] so they could see how much time they had to kill. Only by the middle of the twentieth century did the commercial market for radium fall away, the evidence by now too compelling[⑪] to ignore; the deadly after-effects of radioactive poisoning too serious to risk. As a medicine, a toy, a gadget and a drink, radium had had its day.

注释

① culminate [ˈkʌlmɪˌneɪt] *v.* 以……告终，结果成为
② mount [maʊnt] *v.* 增强
③ carat [ˈkærət] *n.* 克拉，黄金纯度
④ pouch [paʊtʃ] *n.* 小袋，育儿袋
⑤ condom [ˈkɒndɒm] *n.* 安全套
⑥ velvet-lined [ˈvelvɪt laɪnd] *adj.* 天鹅绒衬里的
⑦ respirator [ˈrɛspəˌreɪtə] *n.* 呼吸机，人工呼吸器
⑧ barbed [bɑːbd] *adj.*（评论或玩笑）带刺的，暗讽的
⑨ trenches [ˈtrɛntʃɪz] *n.*（尤指第一次世界大战时的）战壕
⑩ wristwatch [ˈrɪstˌwɒtʃ] *n.* 手表
⑪ compelling [kəmˈpɛlɪŋ] *adj.* 令人信服的

然而，镭是一种药的观点在人们心中已经根深蒂固。在20世纪30年代，随着越来越多骇人故事和证据的出现，镭水变得不再那么畅销。然而，另一些新的镭产品又出现了。150美元可以买到一个14克拉的黄金Radiendocrinator，在它天鹅绒内衬皮套里装有男士内裤或避孕套。比如，比较便宜的鼻锥呼吸器。对于准备参加第二次世界大战的士兵来说，能在黑暗中发光的镭产品绝对具有吸引力：有放在铁丝网上帮助战友找到通道的镭大头钉，有能在夜间瞄准的镭枪，还有让战壕里的士兵知道离战斗还有多久的夜光手表。直到20世纪中期，镭的商业市场才逐渐消失，在有力的证据下，人们都已知道放射性物质的危害，没人敢去尝试放射性中毒的致死后果。由镭制成的药品、玩具、装置和饮料终于寿终正寝了。

REJECTED

THE 'CLOUDBUSTER'

Austrian psychoanalyst[1] Wilhem Reich's plans to form clouds using a type of radiation he had identified, creating rain along the way, led to at least one bunch[2] of happy farmers being relieved from drought[3]. But 'cloudbusting' was just one of his ideas that led to his life's work – at one stage widely admired – eventually becoming much ridiculed[4].

Combining the looks of a space-age weapon with a rather cumbersome[5] clothesline, Reich's 1940s' cloudbusters fired streams of potent 'orgone[6]' energy that he claimed surrounded the earth. Any resulting rain was but a by-product of his main intention – which was to harness[7] the sexual orgone energy and use it for the good of mankind. These were worthy aspirations from an intelligent scientist who ended his days certified and imprisoned[8], but whose 'surrealistic[9] creations' (in the words of Psychosomatic Medicine) never managed to produce the results he expected. In keeping with psychoanalytic theory – an emerging discipline in which Reich was intimately[10] involved – the root of the cloudbusters' failure

注释

① psychoanalyst [ˌsaɪkəʊ'ænəlɪst] *n.* 精神分析学家
② bunch [bʌntʃ] *n.* 伙
③ drought [draʊt] *n.* 干旱
④ ridicule ['rɪdɪˌkjuːl] *v.* 嘲笑
⑤ cumbersome ['kʌmbəsəm] *adj.* 笨重的
⑥ orgone ['ɔːgəʊn] *n.* 生命力
⑦ harness ['hɑːnɪs] *v.* 给……套上挽具
⑧ imprison [ɪm'prɪzən] *v.* 监禁
⑨ surrealistic [səˌriːə'lɪstɪk] *adj.*（方法）超现实主义的
⑩ intimate ['ɪntɪmət] *adj.* 亲密的

破云器

奥地利精神分析学家威廉·赖希（Wilhelm Reich）打算通过一种他发现的辐射能来形成云，并产生降雨。因为雨水缓解了干旱，所以许多农民对此感到非常高兴。不过，破云器仅仅是他毕生从事的事业中的一项发明，它曾经为许多人所称赞，最终却备受奚落。

赖希在20世纪40年代发明的破云器看上去就像是把一件太空时代的武器连接到笨重的晾衣绳上一样，用于发射出一种高强度的“活力”（Orgonon）能量流。据赖希说，这种能量流存在于地球周围。他的主要目的是控制性活力能量，用于为人类造福，降雨只是其中的一个副产品。这只是一位才华横溢的科学家令人尊敬的理想抱负，但他的“超现实主义创造”（引自《身心医学》杂志）未能产生他想要的结果，而他最终也被诊断为疯子并遭到监禁。根据精神分析理论（赖希非常熟悉的一门新兴学科），破云器的失败根源在于其发明者的童年。

先是赖希的母亲自杀，接着他父亲深陷丧妻之痛之

lay in its inventor's childhood.

First Reich's mother committed suicide. Then his father, bereft at his wife's death, killed himself by standing for hours in cold water and contracting pneumonia①. Reich's own attitude to the sex and the body was somewhat unusual. His mother had enjoyed a passionate affair with his tutor, which he'd worried about being forced to participate in.

With such intense interest in sex from an early age, Reich set out to dedicate his life to the study of the energy of orgasms. As one of a coterie② of scientists working alongside Sigmund Freud, the psychoanalyst who postulated③ the concept of the 'id', 'ego', and 'superego' at the beginning of the twentieth century, his early career included extensive work linking the body and the mind in a way that is largely accepted today. Had he stopped there, he may have been remembered for his contribution to science. As it is, with the passing of the years, Reich became progressively more eccentric and his opinions increasingly extreme. His sex-political units, established in Austria to psychoanalyse the fervent④ political and economic environment (what he also called sex-economics), were understandably popular.

By now ridiculed in central Europe, Reich left for Norway where he 'discovered' 'bions⑤' – blue organisms that destroyed bacteria and existed in a state somewhere between living and dead matter. This zombie⑥ life force caused skin to tan and emitted orgone radiation: 'the pure energy of life, the raw power of orgasm'. The press loved the story, but the scientific community scoffed⑦. Reich migrated once more, this time heading for America where he began to build 'orgone accumulators⑧', boxes the size of telephone booths in which patients would sit and feel the force of orgone. The aim was medical. Accumulators directed orgone energy – 'the cosmic

注释

① pneumonia [njuː'məʊnɪə] *n.* 肺炎
② coterie ['kəʊtərɪ] *n.* 小圈子
③ postulate ['pɒstjʊleɪt] *v.* 假定
④ fervent ['fɜːvənt] *adj.* 热情的，热诚的
⑤ bion ['baiɔn] *n.* [生态] 生物个体，生物型
⑥ zombie ['zɒmbɪ] *n.* 僵尸
⑦ scoff [skɒf] *v.* 嘲笑
⑧ accumulator [ə'kjuːmjʊˌleɪtə] *n.* 累加器

中，站在冷水中数小时后得了肺炎去世。赖希本人对待性和身体的态度也很不寻常。他的母亲与家庭教师之间发生过一段婚外恋，赖希害怕被卷入其中。

赖希从小就对性怀有强烈兴趣，把他的一生投入到对性高潮能量的研究之中。他是**西格蒙德·弗洛伊德**（Sigmund Freud）身边一小群科学家圈子中的一员。精神分析学家弗洛伊德在20世纪早期曾提出“本我”“自我”和“超我”的概念。赖希早期对身体和心灵的联系进行了广泛研究，这些研究大多在今天都能为人所接受。他要是就此打住，或许他的科学成就能为世人所铭记。然而，随着时间的推移，赖希变得越来越古怪，思想也越来越极端。他在奥地利创立了性—政治学理论，对当时紧张的政治和经济环境进行了精神分析（他也称这个理论为性—经济学）。可以理解，他的这个理论当时颇为流行。

此时，赖希的理论在中欧受到奚落，于是他前往挪威，在那里他“发现”了“生命单元”——一种能杀死细菌的蓝色有机体，它存在于一种介于活物与死物之间的状态。这种丧尸似的活力能使皮肤晒黑，释放出活力辐射能。这种说法受到媒体的欢迎，却遭到学术界的嘲笑。赖希再次移居他国，这次到了美国，在那里他开始制作一种用来治病的“活力收集器”。这是一种电话亭大小的柜子，病人可以坐在里面，感受活力能量。收集器引导活力能量（当时赖希称为“宇宙能量”）用于控制疾病。

新的国家，新的开始，媒体这次偏向赖希，但只

注释

西格蒙德·弗洛伊德（1856—1939），知名精神分析学家，精神分析学的创始人。他提出“潜意识”“自我”“本我”“超我”等概念，著有《梦的解析》《精神分析引论》等著作。

注释

① intensify [ɪn'tɛnsɪˌfaɪ] *v.* 加强，强化
② antidote ['æntɪˌdəʊt] *n.* 解毒药
③ arid ['ærɪd] *adj.* 干旱的
④ apparatus [ˌæpə'reɪtəs] *n.* 设备
⑤ saucer ['sɔːsə] *n.* 茶碟

energy' as Reich was now defining it – to control disease.

With a fresh start in a new country, press opinion turned back in Reich's favour – for a while. Coming soon after his sensible studies of the link between body and mind, and with an American press that hadn't been subjected to his theories on sex-economics, orgone was initially received with appropriately respectful interest. Science and Society recommended his theories as something to watch. As work intensified[①] in 1942, Reich moved his family to a large retreat in Maine that he renamed Orgonon in honour of his discovery. From now on, his inventions focused almost exclusively on orgone, but tests didn't always go to plan. 'Daddy put a radium needle in the big accumulator in the lab and everyone got sick,' said his son Peter. 'The lab closed, the mice died.'

But life was fun for a boy with an eccentric inventor for a dad, particularly when the cloudbuster rolled out of the lab. With its long, parallel tubes set in a frame and pointing at the sky, this was no toy. Anti-orgone, or Deadly Orgone Radiation (produced by atomic testing and UFOs), was changing the climate and the cloudbuster was the antidote[②]. By absorbing orgone through hoses from the machine's pipe that dipped into water, precipitation could be produced. When fired correctly, streams of orgone would form clouds by creating a stronger orgone energy field than that in the atmosphere, from which the orgone would then be sucked down to earth and the resulting energy put to good use.

With the machine operational, Reich looked around the USA for a place to form clouds. Heading for arid[③] Arizona, he packed a cloudbuster and son Peter into a truck and wound his way west. Aiming the apparatus[④] at a clear blue sky, Peter fired. Not long after, it rained – at least according to Peter, who also used the cloudbuster to chase flying saucers[⑤] of green and red

维持了一阵子。那时美国媒体还不知道他的性—经济学理论，而活力作为赖希对身体和心灵之间联系的合理研究结果，在刚开始时得到了美国人应有的尊重和关注。《科学与社会》杂志介绍了他的理论，认为他的理论值得关注。1942年，随着工作的开展，赖希一家迁到缅因州的一处面积很大的地方，为纪念他的发现，他把这个地方改名为Orgonon。从那时起，他的所有发明几乎都与活力有关，但试验总是达不到预期效果。“爸爸把一根镭针放到实验室的那个大收集器里后，每个人都开始觉得不舒服，”他的儿子彼得说，“最后实验室关了门，老鼠也死了”。

不过，对于一个拥有古怪发明家父亲的男孩来说，生活充满了乐趣，特别是当破云器第一次被推出实验室的时候。破云器长长的、平行的管子安装在一个框架内，指向天空。当然，这可不是什么玩具。赖希认为，反活力能量或死亡活力辐射能（由原子弹试验和不明飞行物产生）正在改变气候，而破云器可以抵抗这种负能量。通过浸入水中的机器上的软管吸收活力，就能产生降水。当发射正确时，活力能量流会形成云，形成一个比大气更强的活力能量场，随后活力能量就会被吸到地球上，产生的能量可供人们使用。

破云器做好了，赖希需要在美国找到一个合适的地点来做试验，最后他选定了干旱的亚利桑那州。他带上破云器和儿子彼得，开着一辆卡车一路向西行驶。最后，彼得把这台装置指向湛蓝的天空，开始发射。没过多久就下雨了——至少彼得是这样说的，他还说他用破云器进行了“一次宇宙探险”，追逐绿色和红色的飞碟。

disks: 'a cosmic[①] adventure'.

For a man who was afraid of thunder and lightning, manipulating[②] clouds was bravery indeed. And, while the cloudbuster was specifically intended to form clouds to bring orgone down to earth, a group of worried farmers from Maine hoped it could be the solution to a lengthy drought. On 6 July 1953, Reich was called in to save the state's blueberry crop. Local newspaper the Bangor Daily News reported that just hours after the cloudbuster went into action, the wind direction changed and rain fell. With blueberries saved, happy farmers paid Reich.

But it was the 'orgone energy accumulator' that was to do it for Reich. Believing that orgone-accumulating boxes could do much to reduce pain and disease, he began commercial production. About 250 accumulators went onto the market. Customers liked the idea of being cured of colds or impotency[③] simply by sitting in a box and letting orgone accumulate, but the US Food and Drug Administration were far from impressed with its medical claims. A ban on distributing the product was issued, infuriating[④] Reich. He was not to be stopped and sales continued. Even Albert Einstein is said to have had a go in an accumulator in 1941 when he paid Reich a visit. The Food and Drug administrators, though, remained unmoved. Reich's claims, they said, were fraudulent[⑤], and the case came to court. During his trial for contempt[⑥], Reich, who conducted his own defence, sent the judge all of his books, but the verdict[⑦], when it came, went against him. Reich, jailed for two years, his books destroyed and his reputation ruined, was finished. Showing no willingness to accept the authority of the court, he said he would carry on selling orgone accumulators regardless: thus dashing[⑧] any hopes he may have harboured for a suspended sentence. Paranoia[⑨] and delusions[⑩] of grandeur were diagnosed as he

注释

① cosmic ['kɒzmɪk] *adj.* 宇宙的
② manipulate [mə'nɪpjʊˌleɪt] *v.* 控制
③ impotency ['ɪmpətənsɪ] *n.* 无力，无效
④ infuriating [ɪn'fjʊərɪˌeɪtɪŋ] *adj.* 激怒人的
⑤ fraudulent ['frɔːdjʊlənt] *adj.* 欺骗性的
⑥ contempt [kən'tɛmpt] *n.* 蔑视
⑦ verdict ['vɜːdɪkt] *n.* 裁定，（陪审团的）裁决
⑧ dashing ['dæʃɪŋ] *adj.* 时髦的
⑨ paranoia [ˌpærə'nɔɪə] *n.* 多疑，恐惧
⑩ delusion [dɪ'luːʒən] *n.* 错觉

对于一个害怕打雷和闪电的人来说，要控制云的确是一件需要勇气的事。虽然破云器专门用于形成云，把活力能量带到地上，但缅因州的一群农民却希望它能解决长久干旱问题。1953年7月6日，赖希应邀去挽救该州的蓝莓。根据当地报纸《班戈每日新闻报》报道，就在破云器发射后几个小时，风向变了，开始下雨了。蓝莓得救了，农民们非常高兴，给了赖希一些报酬。

对于赖希来说，他真正要做的是“活力能量收集器”。他坚信这种收集活力能量的柜子有助于缓解病痛，于是开始商业化生产这种收集器，约向市场投放了250台收集器。病人只需坐在一个柜子里，等待活力聚集。顾客倒是挺喜欢这种治疗感冒或性无能的方法，不过美国食品药品监督管理局却对此不太感冒，禁止销售这种产品。赖希被激怒了，他绝不会就此放弃，仍然继续销售这种机器。据说甚至爱因斯坦在1941年拜访赖希时也试过这种收集器。尽管如此，美国食品药品监督管理局仍然禁止销售收集器，他们说赖希的说法属于欺诈行为，还把他告上了法庭。在审判期间，为自己辩护的赖希把自己所有的书都送给了法官，可最后还是败诉了。他被判处两年徒刑，他的书被销毁。他名誉扫地，再无出头之日。可是，他没有表现出愿意接受法庭的裁决，还说将义无反顾地继续销售活力收集器，这使他获取缓刑的希望也破灭了。他在坐牢时，被诊断为患有偏执和妄想自大症，在监狱里也未能有所好转。

赖希60岁那年，再过两个月就要获释时，却因心脏病发作而去世，而把活力作为一种有价值、严肃的科学研

went to prison. And it didn't get any better from there.

Aged 60 and just two months from release, a heart attack killed Reich, and with him died any hopes of establishing orgone as worthy of serious scientific study. He passed away believing that he was the victim of both a communist[①] conspiracy[②] and a cosmic war. Buried at Organon in a coffin[③] he'd bought a year before his death, a replica[④] of a cloudbuster stands by his graveside[⑤]. His work is immortalised[⑥] too in singer Kate Bush's album Cloudbuster, which also carries an illustration[⑦] of the device.

注释

① communist ['kɒmjʊnɪst] *n.* 共产主义者
② conspiracy [kən'spɪrəsɪ] *n.* 合谋
③ coffin ['kɒfɪn] *n.* 棺材
④ replica ['rɛplɪkə] *n.*（雕像、建筑物或武器等的）复制品
⑤ graveside ['greɪv͵saɪd] *n.* 坟边
⑥ immortalise [i'mɔ:təlaiz] *v.* 使不朽
⑦ illustration [͵ɪlə'streɪʃən] *n.*（书中的）插图

究的希望也随他而去。他离世前还坚信，自己是宇宙大战的牺牲品。他被葬在Orgonon他死前一年买好的棺材里，一件破云器的复制品立在他的墓边。他的发明在歌手凯特·布什（Kate Bush）的专辑《破云器》中被歌颂，专辑中还有破云器的图例。

ABANDONED

ESCAPE COFFINS FOR THE MISTAKENLY INTERRED

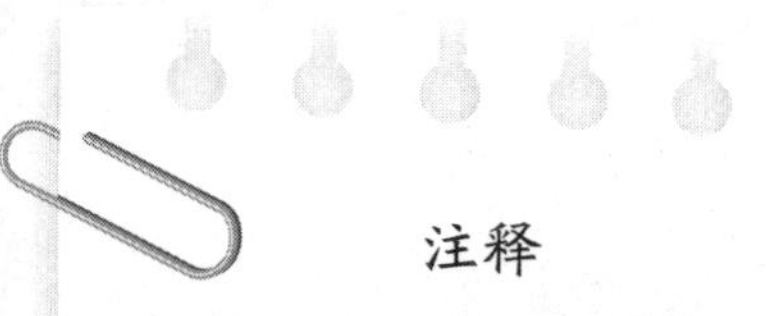

注释

① nuisance ['nju:səns] *n.* 令人讨厌的人或事物，麻烦的人或事情

② terminal ['tɜ:mɪnəl] *adj.*（疾病）晚期的，致命的

③ Taphophobics [ˌtæfəʊ'fəʊbɪks] *n.*（来自希腊）活埋恐惧症患者

④ cataleptic [ˌkætə'lɛptɪk] *n.* 全身僵硬症病人

⑤ paralysis [pə'rælɪsɪs] *n.* 瘫痪

⑥ contraption [kən'træpʃən] *n.* 古怪装置

⑦ salvation [sæl'veɪʃən] *n.* 解救，拯救

⑧ resuscitate [rɪ'sʌsɪˌteɪt] *v.* 救醒

⑨ raging ['reɪdʒɪŋ] *adj.* 激烈的，强烈的

Death is always a nuisance①, but failing to die before burial can prove terminal②. In the eighteenth and nineteenth centuries the thought of waking trapped in a coffin, deep in a grave, was a very real fear, and, evidence suggests, sometimes a reality. Many are the tales of those prematurely buried, sometimes deliberately as a form of execution, but sometimes accidentally too.

So what could be handier than a coffin that those prematurely interred could climb out of ? Taphophobics③, those who fear being buried alive, and cataleptics④, those suffering paralysis⑤ so severe it gives the impression of death, could rest easy with a range of contraptions⑥ to bring salvation⑦.

For at that time, death suddenly wasn't always what it seemed. What had previously been a cut and dried matter of the heartbeat stopping was no longer necessarily so. Medical advances, slow though they were, meant that people who up until that point were formally deceased could now sometimes be resuscitated⑧. Heartbeats could return. And the debate raging⑨ in medicine, in philosophy and in literature, frightened people to death. In the 1730s Jacques Benigne Winslow captured the

避免误埋的逃生棺材

死亡总是一件麻烦事，但在埋之前还没死那就真是麻烦到极点了。想想看，在18世纪和19世纪，那些深埋在墓穴棺材里的人醒来时发现自己被困在里面，真是令人毛骨悚然。而且有证据表明有时真会出现这种情况，很多时候是埋早了，有时候是作为一种死刑故意活埋的，但有时候则纯属意外。

那么，有什么比让那些被过早埋藏的人能爬出来的棺材更方便的呢？活埋恐惧症患者（即害怕被活埋的人）和全身僵硬症患者（即麻痹严重到让人感觉到已经死亡的人）可以从容地躺在一些奇妙的装置中等待超度。

因为在那个年代，突然死亡往往不是表面上看起来那么回事。以前只凭心跳停止就能判断死亡，现在不是这样了。医学进步尽管非常缓慢，但现在即使心跳停止的人有时也可以被救活，心脏可以再次跳起来。反倒是医学界、哲学界、文学界展开的激烈辩论能把人们吓死。18世纪30年代，雅克·贝尼涅·温斯洛（Jacques Benigne

注释

① eloquent ['ɛləkwənt] *adj.* 雄辩的，有说服力的
② precipitate [prɪ'sɪpɪteɪt] *adj.* 仓促的
③ interment [ɪn'tɜːmənt] *n.* 坟墓
④ dissection [dɪ'sekʃn] *n.* 解剖
⑤ decay [dɪ'keɪ] *n.* 腐坏
⑥ corpse [kɔːps] *n.* 尸体
⑦ putridity [pjuː'triditi] *n.* 腐败，堕落
⑧ scurvy ['skɜːvɪ] *n.* 坏血病
⑨ eternity [ɪ'tɜːnɪtɪ] *n.* 永恒，永存
⑩ eventuality [ɪˌvɛntʃʊ'ælɪtɪ] *n.* 可能发生的（尤指不测）事件，可能的后果
⑪ stiff [stɪf] *adj.* 硬的，不易弯曲的
⑫ mourner ['mɔːnə] *n.* 送葬者，哀悼者
⑬ extinct [ɪk'stɪŋkt] *adj.* 不再存在的
⑭ sliding ['slaɪdɪŋ] *adj.* 根据给定规格起降的

spirit eloquently[①] in the title of his book *The Uncertainty of the Signs of Death and the Danger of Precipitate[②] Interments[③] and Dissections[④]*, which related how common it was for death to be declared prematurely. Even decay[⑤] of the supposed corpse[⑥] couldn't be taken as a sure sign of death, according to researcher Charles Kite, since putridity[⑦] was also a symptom of advanced scurvy[⑧]. Some accounts suggested that one in ten burials may have been conducted a little hastily. And so the inventions came.

Franz Vester's 'improved burial case' – what most people, dead or alive, would know as a coffin or casket – became the first US patent of its type in 1868. Vester's 'improvement' was primarily a wide vertical tube running up 6ft of earth to the ground above the grave, attached at a 90° angle to the box in which one usually spends eternity[⑨]. On waking, relieved to be alive and to have had the foresight to prepare for such an eventuality[⑩], the corpse would then use a ladder provided in the lining of the tube to ascend to ground level. Anticipating some difficulties in manoeuvring oneself around the coffin, presumably a little stiff[⑪] from lying around to be able to use the ladder easily, Vester supplied back-up in the form of a bell. By tugging on a cord placed in the hands before interment, the corpse would ring to attract attention in the graveyard above, no doubt to the alarm of mourners[⑫] at neighbouring graves.

Vester admitted that, in most eventualities, neither the ladder nor the bell would see active service and that, after a suitable period of mourning, people should accept the inevitable. With an early eye on recycling possibilities, he wrote: 'If, on inspection, life is extinct[⑬], the tube is withdrawn, the sliding[⑭] door closed, and the tube used for a similar purpose.' Of its type, Vester's improved burial case possessed a remarkable level of ingenuity not considered by lesser inventors. Several designs included an alarm system of some kind, a bell on the surface being the most

Winslow）在《死亡标志的不确定性》和《仓促埋葬与解剖的危害》两本书中叙述了他论点中的精华。这两本书都阐述了过早宣布死亡在当时是非常普通的现象。根据研究员查尔斯·凯特（Charles Kite）的研究，即使认定死亡的尸体腐烂了也不能作为死亡的标志，因为腐烂同时也是坏血病晚期的症状。据说每十个下葬的人里就有一个被埋得有点儿早。于是这个发明应运而生。

弗朗茨·韦斯特（Franz Vester）发明的改进型埋葬容器——大部分人，不管活人还是死人，都知道那是一种棺材——于1868年成了美国的第一例棺材专利。弗朗茨的"改进"主要是将一根竖直的宽管子通到墓穴上方，管长约1.8米（6英尺），与棺材呈90度角。"死者"一旦醒来，发现自己还活着，由于预料到有这种可能性，就可以沿管子里的梯子爬到地面上来。韦斯特预料到人在棺材里可能不容易活动，大概因为久躺导致身体有点儿僵硬，不太容易使用梯子，于是又设计了响铃作为备用设备。"死者"可以拉动埋

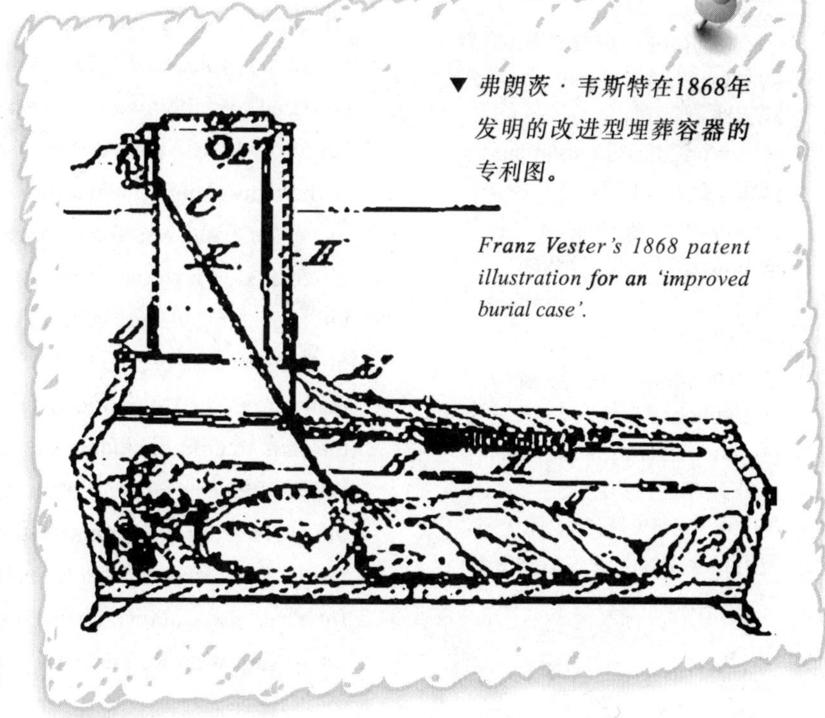

▼ *弗朗茨·韦斯特在1868年发明的改进型埋葬容器的专利图。*

Franz Vester's 1868 patent illustration for an 'improved burial case'.

usual, but they neglected any way of getting oxygen to the body. Even those who awoke to find themselves peering[①] dismally[②] at a coffin lid, but relieved that at least they could ring for service, would be long dead before help came along.

Before Vester's improved burial case, in 1843 Christian Eisenbrandt registered a coffin-dodging device with a spring-loaded[③] lid[④]. But unlike Vester's pioneering idea, it wouldn't function after burial, which was rather a disappointment if you woke up to the sound of earth being shovelled[⑤] on top. 'The slightest motion of either the head or hand acting upon a system of springs and levers cause the instantaneous[⑥] opening of the coffin lid,' says Eisenbrandt's patent application. More Jack in the Box than coffin, there is no record of the device, or Vester's, ever going into manufacture.

So to the grave people continued to go, where presumably they sometimes woke up blinking[⑦] into the darkness and sensing they might have a problem on their hands. In 1858, the magazine Notes and Queries reported how the relatives of a deceased[⑧] wealthy widow, bearing her to interment next to her beloved who had passed away fifteen years earlier, had a shock at the mausoleum[⑨]. When the tomb was re-opened 'the coffin of her husband was found open and empty, and the skeleton discovered in a corner of the vault in a sitting posture'. Around the same time, Edgar Allen Poe, eager to repeat the success of his gothic story The Fall of the House of Usher in which a man buries his cataleptic sister, rushed out more yarns along the same theme, including the self-explanatory The Premature Burial. Readers lapped[⑩] it up, and even the famous, sane and otherwise rational began to take precautions. US President Washington decreed[⑪] that his body should be kept above ground for three days before burial, just in case. And impressionist artist Augustine Renoir, who wasn't to die until 1919, was so

注释

① peer [pɪə] *v.* 费力地看，盯着
② dismally ['dɪzməli] *adv.* 沉闷地，阴暗地
③ spring-loaded ['spriŋˌləudid] *adj.* 弹簧承载的，弹顶的
④ lid [lɪd] *n.* 盖子
⑤ shovel ['ʃʌvəl] *v.* 用铲挖，铲起
⑥ instantaneous [ˌɪnstən'teɪnɪəs] *adj.* 即刻的
⑦ blink [blɪŋk] *v.* 眨（眼睛）
⑧ deceased [dɪ'siːst] *adj.* 去世的
⑨ mausoleum [ˌmɔːsə'lɪəm] *n.*（名人或富人的）陵墓
⑩ lap [læp] *v.* 轻拍，比……领先一圈 lap up 照单全收，欣然接受
⑪ decree [dɪ'kriː] *v.* 发布命令

葬前放在手里的绳子，铃响会引起墓地上方人们的注意，当然也会使在相邻墓地的哀悼者恐慌。

韦斯特承认，在大多数情况下，不论梯子还是响铃都不会被经常用到，适当的哀悼期过后，人们应该接受人死不能复生的事实。韦斯特早就预见到了管子循环利用的可能性，写道："如果经检验没有生命迹象，则收回管子，关闭滑盖，还能把管子用于类似用途。"他的这种改进型埋葬容器非常精巧，超出很多发明家的想象。其他人的几款设计也含有某种警报系统（通常是在地面上安装响铃），但他们没有考虑到如何给棺材内送氧气。即使里面的人有幸醒来，发现自己正凄凉地盯着棺材盖，正庆幸还能拉动响铃请人帮忙时，却可能等不到人来帮忙就一命呜呼了。

在韦斯特发明改进型棺材之前，克里斯蒂安·艾森布兰德（Christian Eisenbrandt）于1843年注册了一种采用弹簧棺材盖的棺材逃生装置。但这种装置不如韦斯特的先进，埋下去之后就不能发挥作用了，如果"死者"醒来之后听到上面正在填土，那该有多失望啊！艾森布兰德在专利申请书上写道："头或手轻轻一动，启动弹簧与杠杆系统，即刻可以打开棺材盖。"这简直不是棺材，更像是魔术箱。艾森布兰德和韦斯特发明的装置都没有投入生产的记录。

于是，人们继续走进坟墓，在那里他们有时可能醒来，在黑暗中眨着眼睛，感觉到自己可能遇到了麻烦。1858年，《备忘和查询》杂志报道过一个有钱寡妇的亲戚们在她死后把她送往15年前去世的丈夫的墓地旁埋葬时，

afraid of premature burial that he had his son instruct doctors to 'do whatever was necessary' to ensure that he was truly dead. But records are sparse[1] to the point of non-existence on the successful deployment[2] of an escapology[3] coffin by anyone formerly declared deceased.

注释

① sparse [spɑːs] *adj.* 稀疏的
② deployment [dɪ'plɔɪmənt] *n.* 部署
③ escapology [ˌɛske'pɑlədʒi] *n.* 脱逃术

他们是多么震惊。当他们把墓穴再次打开时，“发现她丈夫的棺材盖开着，里面空空的，而在墓穴的一角发现了一具坐着的骷髅”。大约在同一时期，埃德加·爱伦·坡（Edgar Allen Poe）渴望再现他的哥特式小说《厄舍古屋的倒塌》的成功。在这部小说中，一个男人活埋了他患有全身僵硬症的妹妹。于是他又赶写了几部同一题材的小说，其中包括冒险解谜小说《过早埋葬》。读者欣然阅读这部小说，甚至名人、正常人和理性的人也都开始采取预防措施。美国总统华盛顿（Washington）下令说，为了以防万一，他死后尸体要在地面停放三天才能下葬。印象派大师奥古斯汀·雷诺阿（Augustine Renoir）1919年去世，之前因为害怕过早下葬，他让儿子告诉医生“采取一切必要措施”确认他已真正死亡。不过，几乎没有关于被提早宣布死亡的人成功运用这种逃生棺材逃生的记录。

CANCELLED
CHADWICK'S MIASMA-TERMINATING TOWERS

注释

① cholera [ˈkɒlərə] *n.* 霍乱
② diphtheria [dɪpˈθɪərɪə] *n.* 白喉
③ typhoid [ˈtaɪfɔɪd] *n.* 伤寒
④ lethal [ˈliːθəl] *adj.* 致命的
⑤ eradicate [ɪˈrædɪˌkeɪt] *v.* 根除
⑥ odour [ˈəʊdə] *n.* 独特气味
⑦ campaigner [kæmˈpeɪnə] *n.* 从事活动者
⑧ credentials [krɪˈdɛnʃəlz] *n.*（表明某人有资格做某事的）资历
⑨ miasma [mɪˈæzmə] *n.* 瘴气
⑩ blight [blaɪt] *v.* 使损害，使（地区）遭殃

Cities kill; especially nineteenth century cities, and most especially smelly cities. When people died in their tens of thousands in urban centres from fearful diseases such as cholera①, diphtheria② and typhoid③, the logic was obvious. Smells caused disease which caused death. So if just one thing could be done to stem the rise of lethal④ diseases in Victorian Britain, nothing would beat eradicating⑤ odours⑥, and Edwin Chadwick, leading health campaigner⑦ and early adopter of the comb-over, was the person to do it.

Chadwick's credentials⑧ for establishing the systems that would improve air quality were outstanding. Unfortunately, in respect of the epidemics striking urban Britain, his assumptions, together with those of many eminent experts in public health, were entirely wrong. Miasma⑨ theory – that air filled with killer particles from decomposing matter caused disease – was broadly accepted in Victorian Britain and Chadwick was one of its greatest proponents.

As he approached his 90th birthday and, as it turns out, his death in 1890, Chadwick formed the Pure Air Company. This gave rise to the exciting prospect of a capital blighted⑩ by

项目取消

查德威克的臭气终结塔

19世纪的城市，尤其是臭气熏天的城市真是要命。市中心有成千上万的人死于像霍乱、白喉、伤寒这样的恐怖疾病，因此城市致命一说的逻辑显而易见。臭气引发疾病，疾病导致死亡。在维多利亚时期的英国，如果只做一件事就能根治当时的致命疾病，那就是根除臭气，而做这件事的人就是主要的卫生倡导者和较早采用秃顶偏分发型的人——埃德温·查德威克（Edwin Chadwick）。

▲ 公共卫生改革家埃德温·查德威克。他在19世纪提出在伦敦建造一些巨塔，吸收洁净新鲜空气并释放到街面的想法。

Public health reformer Edwin Chadwick. In the nineteenth century he proposed construction of huge towers in London to suck cleansing, fresh air down to street level.

查德威克拥有过人的资历去建立改善空气质量的系统。不幸的是，他和那些公共卫生领域的专家关于正袭击英国城区的流行病的设想是完全错误的。臭气理论——空气中充满物质分解带来的致命粒子导致疾病——在维多利亚时期的英国被广泛接受，查德威克也是这一理论的积极拥护者之一。

查德威克在90岁生日临近的时候，也就是在1890年他去世那年组建了纯净空气公司。这家公司在号称

注释

① belch [bɛltʃ] *v.* 打嗝
② symposium [sɪm'pəʊzɪəm] *n.* 专题研讨会
③ sewage ['suːɪdʒ] *n.*（下水道排出的）废弃物
④ disposal [dɪ'spəʊzəl] *n.* 清除
⑤ prolix ['prəʊlɪks] *adj.*（发言、书籍等）冗长的
⑥ couch [kaʊtʃ] *n.* 长沙发
⑦ strata ['strɑːtə] *n.* stratum 的复数，地层，岩层，社会阶层
⑧ slum [slʌm] *n.* 贫民区
⑨ squalid ['skwɒlɪd] *adj.* 脏乱的
⑩ unsanitary [ʌn'sænɪtərɪ] *adj.* 不卫生的
⑪ proximity [prɒk'sɪmɪtɪ] *n.* 接近
⑫ rot [rɒt] *v.* 腐烂
⑬ corpse [kɔːps] *n.* 尸体
⑭ euphemistically [ˌjufə'mɪstɪklɪ] *adv.* 委婉地，婉言地

smoke-belching[①] industrial chimneys being blessed with what was termed 'London's Eiffel Towers'. Like Edward Watkin and everyone else planning a tall structure at the time, people were keen to associate with the Parisian success. But rather than a mere trifle of entertainment or place to shop, Chadwick's towers would serve a much-needed social benefit; delivering air from on high and distributing it at street level.

Announcing his new enterprise to a Royal Society of Arts symposium[②] that was chewing over issues around sewage[③] disposal[④], one of his specialist subjects, Chadwick explained that miasma could be beaten if sufficiently tall towers were constructed. They would exceed the heights of neighbouring industrial smoke stacks and even the Eiffel Tower itself. The Builder magazine was impressed, if bored. 'Sir Edwin concluded his somewhat prolix[⑤] communication,' it reported, 'by advocating the bringing down of fresh air from a height, by means of such structures as the Eiffel Tower and distributing it, warmed and fresh, in our buildings.' The towers would 'draw down air, by machinery, from the upper couches[⑥] or strata[⑦] of air and distribute it through great cities, like the Metropolis'.

Fresh oxygen could even go directly into slum[⑧] houses, benefiting many families barely surviving in squalid[⑨], unsanitary[⑩] conditions. Industrialisation had led to great numbers of people living in close proximity[⑪], sharing the same air. Many parents not only slept in the same rooms as their children, but often with other families and frequently with, or near, farm animals. Rotten food, rotting[⑫] corpses[⑬], dead animals and sewage were rarely far away. People even kept what they euphemistically[⑭] called 'night soil' that they could sell later for fertilizer. In 1861 the Statistical Society of London visited a single room occupied by five families, four of which ate, sat and slept in a corner each, with the fifth family in the middle. One

"伦敦的埃菲尔铁塔"的保佑之下，给深受浓烟滚滚的工业烟囱危害的首都带来了令人兴奋的前景。那就是有幸拥有了号称"伦敦的埃菲尔铁塔"的纯净空气塔。就像爱德华·沃特金（Edward Watkin）和其他当时计划修建一座高大建筑的人一样，人们热衷于把他们和巴黎的成功联想起来。但查德威克的塔不是只供娱乐或购物的那种地方，而是要带来社会更需要的好处：将空气从高空输送到街道。

查德威克向正在详细讨论污水处理的皇家艺术学会讨论会宣布他的新公司成立，而污水处理是他的专业学科之一。他说建造足够多的高塔，就可以除去臭气。这些高塔的高度将超过相邻的工业烟囱，甚至超过埃菲尔铁塔。《建筑者》杂志对此印象深刻，写道："埃德温爵士总结冗长的交流时，呼吁用埃菲尔铁塔一样的建筑将温暖新鲜的空气从高处输送到我们的房子里。"这些塔将"用机械装置将上层空气输送到像首都伦敦这样的大城市中。"

新鲜氧气甚至可以被直接输送到贫民窟，造福于在肮脏不卫生的条件下勉强生存的众多家庭。工业化使得人口密度增大，太多人呼吸着相同的空气。很多父母不仅和孩子睡在同一间房子里，而且往往是和其他家庭一起，甚至经常是和农场的动物共处或者在它们附近。腐烂的食物、正在腐败的尸体、动物死尸和污水经常就在不远处。人们甚至保存他们委婉称为"夜土"的东西，之后可以当作肥料卖掉。1861年，伦敦统计学会造访了五个家庭居住的一间房屋，四个家庭分别在房子的四角饮食起居，第五个家庭在房子中央。一名妇女告诉调查者："我们本来生活得挺好，中间的那位男士收了一个寄宿者后我们的情况就变

woman told investigators: 'We did very well until the gentleman in the middle took a lodger.'

In 1868, London suffered its Great Stink. It was hardly unexpected. From the 1840s to 1860s, the smell of the capital was overwhelming. Across the country outbreaks of cholera killed tens of thousands of city dwellers. Miasma was blamed. But if Chadwick's plan to suck clean air from the sky to the ground had logic, it was the wrong logic. Against a growing body of evidence – although in line with prevailing opinion – he ignored the criticality of water cleanliness. Instead, The Times reported, Chadwick called for the 'complete drainage[①] and purification of the dwelling house, next of the street and lastly of the river', in effect taking away waste from homes and adding it to the water.

This, though, was the parody[②] of Chadwick's position. He had been a powerful advocate for urban drainage and sewerage systems, and although his contemporary Joseph Bazalgette is seen as London sewers' hero, Chadwick had also campaigned against the degrading and dirty conditions caused by industrialisation. Nevertheless he stuck to his fundamental position on miasma: bad air caused disease. He had high-profile support too. Florence Nightingale, by now famous for her nursing work during the Crimean War, was also a miasma theory advocate; believing that measles, smallpox and scarlet[③] fever were the fault of the new habit of building houses with drains below through which odours could escape and infect the masses. Nightingale proved a much more amenable[④] character than Chadwick and from 1870 onwards had the ear of more people when she talked about sewage.

However, by failing to make the connection between clean water and disease, even though Dr John Snow had discovered this link at about the same time, Chadwick and Nightingale,

注释

① drainage ['dreɪnɪdʒ] *n.* 排水，排水系统

② parody ['pærədɪ] *n.* 滑稽模仿作品（指文章、戏剧、音乐作品的滑稽模仿作）

③ scarlet ['skɑːlɪt] *adj.* 猩红（的），鲜红（的）

④ amenable [ə'miːnəbəl] *adj.* 愿意的

糟糕了。”

1868年，伦敦遭受了“大恶臭”，这一点儿都不意外。19世纪40—60年代，伦敦的臭味势不可挡。全国霍乱暴发导致成千上万名市民死亡。臭气成了众矢之的。但假如查德威克从空中吸取洁净空气输送到地面的计划有逻辑可言的话，那也是错误的逻辑。他与越来越多的证据——尽管他和普遍的意见一致——对阵，无视水清洁的危急程度。相反，《泰晤士报》写道，查德威克号召“彻底排水，净化住房，然后净化街道，最后是河流”，实际上是把垃圾带走，远离房屋然后倒进河里。

然而这只是对查德威克立场的嘲讽。他大力倡导修建城镇污水排水系统，尽管同一时期的约瑟夫·巴扎格特（Joseph Bazalgette）被视为建造伦敦下水道的英雄，查德威克也参与了反对工业化引起的恶劣肮脏环境的活动。但是他坚持关于臭气的根本立场：恶劣的空气导致疾病。他也得到了名人的支持。因在克里米亚战争中出色的护理工作闻名的**弗罗伦斯·南丁格尔**（Florence Nightingale）也支持臭气理论，认为麻疹、天花和猩红热都是由于新的建筑习惯把排水沟建在下面，臭气可以跑出来感染民众而引起的。南丁格尔比查德威克表现出更具责任感的特点，自1870年以来，她每次谈到污水都会引起更多人的重视。

然而，尽管约翰·斯诺（John Snow）博士已经发现了洁净水与疾病的联系，但查德威克和南丁格尔却没有将这两者联系起来，和科学圈的其他人一起仍

注释

弗罗伦斯·南丁格尔（1820—1910）英国护士和统计学家。“南丁格尔”是护士精神的代名词。“5·12”国际护士节设立在南丁格尔的生日这一天，就是为了纪念这位近代护理事业的创始人。

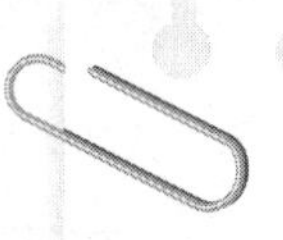

注释

① belch [bɛltʃ] *v.* （大量）喷出，吐出
② grits [grɪts] *n.* 粗碾粉
③ bust [bʌst] *v.* 打碎
④ despise [dɪ'spaɪz] *v.* 鄙视
⑤ forthright ['fɔ:θˌraɪt] *adj.* 直率的
⑥ derogatory [dɪ'rɒgətərɪ] *adj.* 贬低的
⑦ prig [prɪg] *n.* 道学先生，一本正经的人
⑧ mellow ['mɛləʊ] *adj.* 老练的，成熟的

along with most of the rest of the scientific community, were focusing on the wrong cause. Even though Snow's theory was gaining acceptance following a cholera outbreak in an area not covered by London's newly installed sewage system, Chadwick continued to be an exponent of the importance of clean air.

It took many years for miasma theory to be discredited, but even that should not lessen the contribution that both Chadwick and Nightingale made in emphasising the need for clean air. Even as late as the 1950s, poor air quality was dispatching Britain's urban poor in unacceptable numbers. 'It is time that the air we breathe is recognised to be as important as the water we drink,' thundered the British Medical Journal. 'We still treat the air over our large towns as a sewer; into it chimneys belch[①] thousands of tons of harmful grits[②], solids and gases each year. Thus we have failed to learn the lesson of the Thames.'

Chadwick didn't live to see his miasma-busting[③] towers. Nor has anyone else. They had no realistic possibility of gaining acceptance, largely because of Chadwick's age when he conceived them – 90 isn't the ideal time to start building Eiffel Tower-sized structures – but also because of his almost complete lack of interpersonal skills. Universally despised[④] by both rich and poor, he was, according to one biographer: 'The most unpopular single individual in the whole United Kingdom.' As the architect of the workhouse, he was hardly the labouring classes' greatest hero. To make matters worse, he also wanted to restrict the sale of alcohol, one of the few legal substances that made many lives worth living. At the other end of the social scale, colleagues on committees or in his clubs found him quick to express forthright[⑤] opinions in derogatory[⑥] terms. He had the character of 'the bore, the fanatic and the prig[⑦]', said one. The passing of the years did little to mellow[⑧] Chadwick, according to his friend Sir John J. Macdonnell: 'It is one of the few

然把注意力放在错误的起因上。即使在没有被伦敦新建的污水系统覆盖的某地暴发霍乱后，斯诺的理论得到接受，查德威克仍继续坚持洁净空气的重要性。

很多年后臭气理论才受到怀疑，但即使这样也不能减少查德威克和南丁格尔强调需要洁净空气的贡献。甚至到了20世纪50年代，英国大量的城镇贫民仍死于质量差的空气。《英国医学杂志》呼吁“是时候将我们饮用的水和呼吸的空气放在同等重要的位置上了。我们仍然把大片城镇上方的空气当作下水道，烟囱每年向上空喷出成千上万吨有害粉末、硬粒和气体。这样，我们没能从泰晤士河的例子中吸取教训。”

查德威克在有生之年没能看到这些处理臭气的塔，也没有其他任何人看到。这些塔在现实中没有被接受的可能性，很大原因是由于查德威克构思这些塔时的年龄——90岁不是开始修建埃菲尔铁塔一样规模建筑的理想年龄——还因为他几乎完全不懂人际交往的技巧。穷人和富人都看不起他，按照一位传记作者的说法，他是“整个英国最不受欢迎的人”。作为济贫院的建筑师，他压根不是劳动阶级的伟大英雄。更糟糕的是，他还想限制酒类销售，酒类可是让很多人感觉值得活下去的少数合法物质之一。而在社会的另一端，议会或他社团的同事发现他会用贬损的措辞迅速直接地表达意见。有人说他有着“令人讨厌、狂热和一本正经”的特性。过去的岁月没能让查德威克变得柔和一些，据他朋友约翰·J. 麦克唐纳（John J. Macdonnell）先生说：“令人讨厌是老年人享有的为数不多的特权之一，而这个伟人恐怕过早地、过于自由地、过

unquestioned privileges of old age to be a bore, and this great man had, I fear, discounted too early, too freely, and too heavily this privilege. One reason was that he babbled① too much, not of green fields, but of sewage②.'

In the end, Chadwick's Pure Air Company built no towers. Another of his schemes, to train rider-less fire horses to gallop③ automatically to burning buildings whenever they heard a fire alarm, didn't get out of the starting blocks. And to the dismay of generations of children since, he also failed to have spelling tests abolished in schools. They were 'quite unnecessary', he said, particularly as spelling lessons cost two-thirds of the sum spent on elementary education. But, as befits the first president of the Chartered Institute of Environmental Heath, Chadwick did want the Pure Air Company to survive and to provide a social benefit. Reflecting the teachings of his friend and benefactor④ Jeremy Bentham (whom we will meet next), any profit, Chadwick said, would not be large in order that the greatest good could be done for the greatest number of people. He was right in that the Pure Air Company wouldn't make a large profit. In fact, it made none at all, and no pure-air towers either. As he took his last breath of clean, fresh air – he lived in a nice part of town – his exciting project remained unfulfilled.

注释

① babble ['bæbəl] *v.* 含混不清地说，兴奋地说
② sewage ['suːɪdʒ] *n.*（下水道排出的）废物
③ gallop ['gæləp] *v.* 使（马）疾驰，（马）疾驰
④ benefactor ['bɛnɪˌfæktə] *n.* 捐助人

于严重地使用了这种特权。原因之一就是他不是对绿地，而是对污水喋喋不休。”

最终，查德威克的纯净空气公司什么塔都没建成。他的另一个计划——训练无人骑的消防马一听到火警就自动疾驰到起火的房子——也没能启动。此外，令后来的孩子们沮丧的是，他也没能废除学校的拼写测验。他认为："拼写测验非常没有必要"，因为拼写课程占了整个小学教育2/3的时间。但是，为成为称职的特许环境健康研究所第一任所长，查德威克真的希望纯净空气公司能存活下来并为社会带来福利。他对他朋友兼赞助人杰里米·边沁（Jeremy Bentham）的教义进行沉思后，说这个公司的所有利润都不能太大，以便为大多数的人提供最大的好处。事实上，这个公司根本没有利润，也没有造出任何一座纯净空气塔。他在呼吸生命中最后一口洁净新鲜的空气时——他住在城镇里的一个好地方——他的激动人心的愿望仍然没能实现。

FAILED

THE SELF-CLEANING HOUSE

注释

① liberate ['lɪbəˌreɪt] *v.* 解放
② pill [pɪl] *n.* 药丸
③ scratch [skrætʃ] *v.* 挠
④ cranny ['krænɪ] *n.* 缝隙

Frances Gabe hated housework. 'A thankless, unending job,' she called it. But if for much of her lifetime in the twentieth century a woman's place was in the home, she was going to create a home she didn't have to clean. Inspired, thought many. Completely nuts said others. But Mrs Gabe, coming out from a divorce and looking for something to occupy her (other than the housework) did build the world's first, and only, self-cleaning house.

If, as the Vatican decided in 2009, the washing machine did more to liberate① women in the twentieth century than the pill② or the right to work, Mrs Gabe was going to go one better. With sixty-eight mechanisms that take the effort out of cleaning, the whole 30ft × 45ft house at Newberg, Oregon that she built from scratch③ in her 60s is one giant washing machine. At a push of a button, small ceiling-mounted devices in every room would spring into action, running through an entire cleaning-drying-heating-cooling cycle and reaching deep into every nook and cranny④ with no need for anyone to deploy a duster or manhandle a vacuum cleaner. In a passable impression of an industrial car wash, jets of soapy water fired into action first,

自洁式房屋

弗朗西丝·加布（Frances Gabe）讨厌做家务。她把家务称为“吃力不讨好的而且永无止境的工作”。但因为她一生中的大部分时间都生活在20世纪这个女性做家庭主妇的时期，所以她打算发明一种不用打扫的房屋。有人说她是受灵感激发，也有人说她彻底疯了。但是加布夫人离婚后想要找点事情（除了家务以外的事情）打发时间，她真的建成了世界上第一座，也是唯一的一座自洁式房屋。

如果说像罗马教廷2009年判定的那样，在20世纪，洗衣机比避孕药或工作权利更能解放妇女，那么，加布夫人的发明将会更胜一筹。她60多岁时在俄勒冈州的新堡市新建了一座房子，整个约9米×14米（30英尺×45英尺）那么大，有68台机械装置完成清洁任务，这座房屋就是一台巨大的清洗机。一按按钮，安装在每个房间天花板上的那些小型装置就活跃起来，完成整个清洁—烘干—加热—冷却的循环过程，并且能深入每个角落和缝隙，不需要人工动用抹布或吸尘器。通过对工业化的洗车方法的尚属不错

followed by a quick rinse[①] and finally a heated blow dry. The full cycle took forty-five minutes. To ensure no damp patches[②] remained, the floors of the house sloped[③] slightly so that excess water drained away.

The thirty-eight patents for the house and its devices included bookshelves that dusted themselves and contained books with self-cleaning jackets; a fireplace that removed its own ashes; and a bathroom suite comprising of sink, shower, lavatory and bath that squirted[④] themselves with detergent[⑤] and had a good scrub after ablutions[⑥] were complete. In the kitchen, the standard dishwasher was replaced by integral cleaning mechanisms in the cabinets, so that dirty crockery could be stacked[⑦] in its rightful place and washed ready for the next meal. And the washing machine was redundant, as clothes were laundered[⑧] as they hung in the wardrobe. You simply took off your dirty clothes, hung them up and set the cycle. An hour later they were clean and dry, ready to be worn afresh.

There were downsides. Fixtures, fittings and fabrics[⑨] had to be completely waterproof (although none of Mrs Gabe's furniture was of Edison's concrete type). Carpets[⑩] were entirely out of the question, as was wallpaper. 'Nothing but dirt-collectors,' said Mrs Gabe.

Conceived largely in the 1950s, when mod cons were coming into the home and people were beginning to afford new labour-saving devices, the house was lauded[⑪] and laughed at in equal measure, with the Massachusetts Institute of Technology, one of the world's leading centres for innovation and enterprise, declaring Mrs Gabe one of America's leading female inventors. On the sitting room wall she displayed the original patents – some of the world's longest and most complex – safely encased[⑫] in plastic, so guests could admire them as they relaxed on the plastic-wrapped upholstery[⑬].

注释

① rinse [rɪns] *n.* 冲洗
② patch [pætʃ] *n.* 小块
③ slope [sləʊp] *v.*倾斜
④ squirt [skwɜːt] *v.* 挤，喷出
⑤ detergent [dɪ'tɜːdʒənt] *n.* 清洁剂
⑥ ablutions [ə'bluːʃənz] *n.* 洗浴
⑦ stack [stæk] *v.* 堆放，摞起
⑧ launder ['lɔːndə] *v.* 洗
⑨ fabric ['fæbrɪk] *n.* 织物
⑩ carpet ['kɑːpɪt] *n.* 地毯
⑪ laud [lɔːd] *v.* 嘉许，称赞
⑫ encase [ɪn'keɪs] *v.* 包，围
⑬ upholstery [ʌp'həʊlstərɪ] *n.* 座套

的模仿，在洗涤房屋时首先是用肥皂水喷洒，然后迅速冲洗，最后用热风吹干。整个过程需要45分钟。为确保没有湿的地方，房子的地面稍微倾斜，以便多余的水排出。

这座房子及其设备的38项专利包括能自行清除灰尘的书架，并且书架上的书都有能自洁的护套；能自动清除炉灰的壁炉；还有水槽、淋浴、厕所和浴盆构成的浴室组合，在洗浴完成后它们都能自行喷射清洁剂，还能自行擦洗。橱柜中的整体清洗装置代替了厨房内标准的洗碟机，这样，脏餐具就能被堆放在适当的位置加以清洗，以备下次吃饭时再用。洗衣机在这里也是多余的，因为衣服挂在衣柜里时就被清洗了。你只要脱下脏衣服，把它们挂起来，启动程序即可。一个小时后，它们就洗干净并烘干了，以供重新穿着。

这种房屋也有不好的地方。家具、家具配件和纤维织物都必须完全防水（尽管加布夫人的家具都不是爱迪生的混凝土家具那种类型）。地毯和壁纸是绝对不能用的。加布夫人说："那不过是些容易藏污纳垢的东西而已。"

这房子主要是20世纪50年代的构想，那时现代化生活设备进入家庭，人们开始有能力购买新的节省劳力的设备，世界上领先的创新及企业中心——美国麻省理工学院宣布加布夫人是美国主要的女性发明家之一，此时赞美和嘲笑这座房子的人不相上下。她在起居室的墙上展示了最初的专利——有一些是世界上最长最复杂的——用塑料严实地包起来，这样客人们在塑料包装的坐垫上休息时可以欣赏它们。

加布夫人在这座原始的样本房中一直生活到90多岁，

注释

① trample ['træmpəl] *v.* 践踏，踩坏

② ivy ['aɪvɪ] *n.* 常春藤

③ bull [bʊl] *n.* 公牛

Mrs Gabe, who lived in the original prototype of the house well into her 90s, thought that time spent cleaning which instead could be spent with one's family or on self-improvement, was time wasted. She spent twelve years working on the inventions in her garden, where she installed a series of shower stalls to investigate the behaviour of water at different pressures and on different substances. The house, she believed, would appeal to those who found the work/life balance difficult to manage and the elderly or those with disabilities who simply couldn't keep up the running of a home without help. Women, in particular, would admire the self-cleaning house – unless they are cleaners. 'The problem with houses is that they are designed by men,' Mrs Gabe explained. 'They put in far too much space and then you have to take care of it.'

Despite the attractiveness of the idea and the thoroughness with which Mrs Gabe thought of every possible defect, the concept failed to take off commercially and the patents expired. While the cleaning element may sound practical, living in a house bereft of soft furnishings, with slightly sloping floors and furniture that's encased in plastic, clearly isn't everyone's idea of a comfortable home. And things haven't always gone smoothly. Once the ceilings were damaged in a flood, then the house was hit by an earthquake, putting most of the self-cleaning mechanisms out of action. But Mrs Gabe lived a long and happy life at the house, becoming something of a minor celebrity. Even her oddities were celebrated. Door frames were designed deliberately low so that guests had to bow to her when they came in and a sign outside read: 'Please do not trample[①] the poison ivy[②] or feed the bull[③].'

她认为在打扫房子上花费时间就是浪费，这些时间本可以用来陪伴家人或提升自我。她用了12年在自己的花园里进行研究发明，她在那里安装了一系列淋浴室用来观察水在不同压力下和在不同物质上的特性。她相信这种房子会吸引那些难以平衡工作与生活的人、老年人或者没有帮助就没法持家的残疾人。女性尤其会称赞这种自洁式房屋——除非她们是清洁工。加布夫人解释道：“房子的问题在于它们是由男人设计的，他们在房子里留了过多的空间，然后你还得打扫这些地方。”

尽管这个主意很吸引人，而且加布夫人对可能存在的缺点也想得很周到，但这个理念还是没能取得商业上的成功，并且专利也过期了。虽然清洁原理听起来是实用的，但是生活在没有柔软家具、地板稍微倾斜、家具外面裹着塑料的房子里，这显然不是大家所认为的舒适的家。加之生活总不会一帆风顺，一旦房屋漏雨、房子遭遇地震，那么大部分自洁装置就失灵了。但是加布夫人在这座房子里度过了漫长而幸福的一生，成了小有名气的人。就连她的怪癖也出名了——她故意把门框设计得低了一些，这样客人进门的时候都得向她鞠躬。门外牌子上写着：“请不要践踏毒漆藤，也不要喂牛。”

ABANDONED
THE JAW-DROPPING DIET

It is said that nearly 1,000 years ago William the Conqueror, worried about his weight, went on a self-imposed① 'alcohol only' diet. And in the sixteenth century a Venetian man called Luigi Cornaro made something of a name for himself by living to 100 on a diet which appears not to have consisted of much more than eggs and wine. But diet regimes② as the popular fads③ we know today, only really began to emerge in the nineteenth century. At first there was the Reverend Sylvester Graham, the father of Graham crackers④, who promoted simple foods and vegetarianism. Then, in 1863, an undertaker called William Banting wrote what was really the first handbook for dieters, *A Letter on Corpulence*, which became a bestseller. It warned slimmers⑤ off starchy⑥, sugary fare⑦ in favour of lean meat, eggs and vegetables and has been seen as a forerunner of the modern Atkins Diet.

Then, towards the final years of the century, the American Horace Fletcher arrived on the scene to cash in on the new dieting craze. Today few of us have heard of Fletcher and even fewer are following his eating plan which would see us chew each morsel⑧ of food over and over again. Yet in his day

注释

① self-imposed [ˌself ɪm'pəʊzd] *adj*. 自愿承担的，自己强加的

② regime [reɪ'ʒiːm] *n*. 养生法

③ fad [fæd] *n*. 一时的狂热

④ cracker ['krækə] *n*. 薄脆饼干

⑤ slimmer ['slɪmər] *n*. 减肥者，减轻体重者

⑥ starchy ['stɑːtʃɪ] *adj*. 含大量淀粉的

⑦ fare [fɛə] *n*. （尤指餐厅或特殊场合吃的）饭菜

⑧ morsel ['mɔːsəl] *n*.（尤指食物）极少的量

项目废弃

令人目瞪口呆的饮食法

据说1000年前，征服者威廉对自己的体重忧心忡忡，就强迫自己通过“只喝酒”的方法减肥。16世纪时，一位名叫路易吉·科尔纳罗（Luigi Cornaro）的威尼斯人，光靠鸡蛋和葡萄酒就活到100岁，因此而留名青史。但是我们现在知道的时尚饮食养生法也不过在19世纪才真正开始出现。首先是全麦饼干之父西尔维斯特·格雷厄姆牧师（Sylvester Graham），他提倡吃简单的食物和素食。后来殡仪师威廉·班廷（William Banting）在1863年写了《一封关于肥胖的公开信》，这是第一本真正面向节食者的指南，并成了畅销书。书中警告减肥者要远离含淀粉和糖分的食物，提倡食用瘦肉、鸡蛋和蔬菜，这种方法被视为现代阿特金斯减肥法的先驱。

到了19世纪最后几年，美国人贺拉斯·弗莱彻（Horace Fletcher）在新的节食狂潮中捞了一笔。现在很少有人听说过弗莱彻，甚至没人遵循他每口食物都要反复咀嚼的饮食方法。但在他那个时代，他的节食方法获

Fletcher's diet gained mass popularity and advocates included the likes of the oil tycoon[1] John D. Rockefeller and the authors Henry James and Franz Kafka. His chew-chew diet would make him a fortune and earn him the nickname 'The Great Masticator'!

Fletcher was a successful art dealer but by the time he reached his 40s he wasn't happy. He was just 5ft 6in but weighed[2] more than 15 stones and felt, he said later, like 'an old man'. Then, in the early 1890s, he was turned down for life insurance. It was this that led Fletcher to drastic measures. In 1898 he was killing time over a hotel meal on a business trip to Chicago when he got to thinking about eating styles and its effect on health. He began devising an extreme food plan to battle his own bulge[3]. And within just five months of trying it out he had lost more than 60lbs.

So how did it work? Fletcher was inspired by the advice of former British prime minister William Gladstone to his children, to chew their food 32 times, once for each tooth. He made this notion the nub[4] of his new diet. Food, he said, needed to be properly masticated[5] before it was swallowed. In order to achieve this, dieters should munch[6] whatever they were eating to the point where it became completely liquidised. The precise number of chews would depend on the type of food. Bread, for instance, might take as much as seventy chews. A shallot[7] might need as much as 700. The idea was that all the effort of this would slow people down, so that they consumed less in total. If a food couldn't be entirely chewed down into liquid form? Simple – spit[8] it out. As well as stopping you eat too much, Fletcher also thought that his chewing methods would allow the body to absorb the most potential goodness from the food.

Fletcher's first chewing work emerged in 1898 and he went on to publish several works over the next twenty years outlining

注释

① tycoon [taɪ'kuːn] *n.*（工商界的）大亨
② weigh [weɪ] *v.* 重量为
③ bulge [bʌldʒ] *v.* 鼓起
④ nub [nʌb] *n.* 中心，核心
⑤ masticate ['mæstɪˌkeɪt] *v.* 咀嚼，嚼碎
⑥ munch [mʌntʃ] *v.* 大声咀嚼
⑦ shallot [ʃə'lɒt] *n.* 青葱，大葱
⑧ spit [spɪt] *v.* 吐

得了大众的青睐，此法的众多拥护者中包括石油大亨约翰·D. 洛克菲勒（John D. Rockefeller）和亨利·詹姆斯（Henry James）、弗朗茨·卡夫卡（Franz Kafka）两位作家。他的反复咀嚼饮食法使他发了大财，并获得了“伟大的咀嚼者”的称号。

弗莱彻是一位成功的艺术品商人，但他到40多岁时，并不幸福。他身高只有不到1.7米（5.6英尺），但体重却超过了95千克（15英石），他后来说感觉自己“像个老人”。到了19世纪90年代初，他买人寿保险遭到拒绝。就是这件事导致弗莱彻采取了过激方法。1898年，他在去芝加哥出差途中在酒店用餐时消磨时间，开始思考吃饭方式对健康的影响。开始时，他设计了一项极端的饮食计划来与自己臃肿的身体作斗争。试验了仅5个月，他就减掉了约27千克（60多磅）。

那么，这是如何做到的呢？英国前首相威廉·格莱斯顿（William Gladstone）曾经告诫自己的孩子们，食物要嚼32下，相当于每颗牙齿动一下。弗莱彻受到启发，将这一见解作为自己新节食方法的要点。他说，食物在吞咽前要充分咀嚼。为了做到这一点，节食者无论吃什么都要用力咀嚼，直到食物完全液化。具体的咀嚼次数取决于食物种类。例如，面包可能要咀嚼70次，葱可能要咀嚼700次。这样会使人们慢下来，从而总体上就吃得少了。如果遇到不能咀嚼成完全液化的食物怎么办？好办——吐掉就是了。弗莱彻认为他的咀嚼方法除了防止人吃得过多之外，还能使身体从食物中最大化地吸收营养。

弗莱彻的第一部关于咀嚼的著作于1898年问世，他在

注释

① esteemed [ɪ'stiːmd] *adj.* 受尊敬的
② chorus ['kɔːrəs] *n.* 齐声（批评、反对或赞扬）
③ doctrine ['dɒktrɪn] *n.*（尤指宗教的）信条，学说
④ Fletcherism ['flɛtʃəˌrɪzəm] *n.* 细嚼慢咽法，弗莱彻主义（主张对食物细嚼慢咽）
⑤ dub [dʌb] *v.* 把……称为
⑥ cornflake ['kɔːnfleɪk] *n.* 玉米片
⑦ bronchitis [brɒŋ'kaɪtɪs] *n.* 支气管炎
⑧ revolting [rɪ'vəʊltɪŋ] *adj.* 令人厌恶的
⑨ assertion [ə'sɜːʃ(ə)n] *n.* 明确肯定，断言
⑩ ailment ['eɪlmənt] *n.* 小病
⑪ far-fetched ['fɑː'fetʃt] *adj.* 牵强的
⑫ invigorate [ɪn'vɪgəˌreɪt] *v.* 鼓舞，增加活力

his philosophy. In January 1904 the esteemed[①] medical journal *The Lancet* joined a chorus[②] of approval reporting that: 'a more generally beneficial doctrine[③] could hardly be chosen.' Fletcherising[④], as the new chewing diet was dubbed[⑤], became all the rage on both sides of the Atlantic. Muncheon parties even became popular where food would be served and people would chew each bite simultaneously for five minutes before a bell was rung indicating that they could swallow. Fletcher became good friends with J.H. Kellogg, the inventor of cornflakes[⑥], who recommended his methods to clients. Even a section of the British army road tested his methods, though with mixed results.

Most diets, like Fletchers, are fads. They come and go as quickly as the fast food that so often makes them necessary in the first place. And when Fletcher died in 1919, from bronchitis[⑦], his regime was already losing favour as other diet plans appeared. Yet even modern experts believe there's some method in Fletcher's practices. Several scientific studies now show that the longer people take over their dinner the less likely they are to overeat. And along with all that chewing, Horace had some good, if basic, practical advice, such as only eating when you feel hungry. When the historian Sir Roy Strong tested Fletcher's plan and other diets from history he found that, though revolting[⑧], Fletcherising was one of the most effective historical diets in achieving weight loss in his subjects.

Of course Fletcher wasn't a doctor and his methods were largely based on his own personal eating plan. His assertion[⑨] that Fletcherising your food would actually increase your strength and even help cure ailments[⑩] like toothache look far-fetched[⑪] today, especially as he seems not to have properly understood how digestion really worked or the benefits of vitamins and minerals. But Fletcher clearly felt personally invigorated[⑫] by the chewing practice. Aged 58 he supposedly

接下来的20年中又继续出版了几部作品，阐述他的理念。1904年1月，享有盛誉的医学杂志《柳叶刀》也刊登了报道，对他称赞有加："很难再选出更加有益的学说。"这种新的咀嚼饮食法被称为弗莱彻式饮食法，风靡大西洋两岸。午宴聚会也流行开来，食物提供给大家之后，人们每吃一口就会同时一起咀嚼五分钟，然后有铃声提示他们可以咽下了。弗莱彻与J. H. 凯洛格（J.H.Kellogg）成为好友。凯洛格是玉米片的发明者，他把这种饮食方法推荐给了自己的客户。甚至连英国陆军的一个小分队也尝试了这种方法，但是结果有好有坏。

像弗莱彻式咀嚼法这样的饮食法大多是种时尚。它们像快餐一样来去匆匆，其实刚开始往往是快餐让人有了节食的必要。1919年弗莱彻患支气管炎去世时，他的饮食法已风光不再，因为又出现了其他饮食法。不过，就连现代的专家也认为弗莱彻的方法有几分道理。多项科学研究表明，吃饭所用时间越长，人就越不可能吃过量。在用咀嚼法的同时，贺拉斯有一些基本而实用的好建议，例如，只有感觉饿的时候才吃东西。历史学家罗伊·斯特朗爵士（Sir Roy Strong）测试了弗莱彻的饮食法以及历史上出现过的其他方法，他发现，尽管他讨厌细嚼慢咽，但在他的研究对象中，细嚼慢咽是史上饮食法中最有效的减肥方法之一。

当然，弗莱彻不是医生，他的方法主要是基于他的个人饮食计划。他宣称细嚼慢咽真的能使人力气变大，甚至可以帮助治愈像牙疼之类的小病。现在看来，这种说法似乎不靠谱，尤其是他好像还没真正搞明白消化是怎么回事

beat a group of young Yale college athletes in a series of tests at the gymnasium[①] there.

Ultimately, however, his way of eating didn't catch on permanently. Perhaps Fletcher was simply more patient than the general populace[②]. Most people, it turned out, couldn't be bothered to spend all that time chewing. His eating methods certainly might not be all that appealing, especially if you had to share dinner with someone who followed them and were expecting some lively conversation. But they were probably a lot more effective than William the Conqueror's. The king's booze[③]-fuelled[④] weight-loss plan died along with him when in 1087, and still fat, he was mortally wounded after falling off his horse.

注释

① gymnasium [dʒɪm'neɪzɪəm] *n.* 健身房

② populace ['pɒpjʊləs] *n.* 平民

③ booze [bu:z] *n.* 酒

④ fuelled ['fjʊəld] *adj.* 用作燃料的

以及维生素和矿物质的好处。但弗莱彻发现咀嚼法明显使他感到精力充沛了。据说他58岁那年，在当地健身房进行的一系列测试中打败了耶鲁大学的年轻运动员。

然而，他的饮食法最终未能一直流行下去，可能弗莱彻只是比普通人更有耐心吧！事实上，多数人不愿意花那么长时间去咀嚼。他的饮食法也许注定不会那么吸引人，尤其是你得跟一个细嚼慢咽的人一起用餐，而你希望开怀畅聊时。不过，这种饮食法可能比征服者威廉的方法有效得多。这位国王在1087年从马上摔了下来，因伤势过重而去世，去世时还十分肥胖，而他的酒精刺激减肥法也跟他一起消失了。

FAILED

A NUTTY PLAN TO FEED THE MASSES

注释

① equivalent [ɪ'kwɪvələnt] *n.* 等量物，等价物

② millennium [mɪ'lɛnɪəm] *n.* 千禧年 the Millennium Dome 千年穹顶

③ fiasco [fɪ'æskəʊ] *n.* 彻底的失败

④ rein [reɪn] *n.* 执政，掌权

⑤ grandiose ['grændɪˌəʊs] *adj.* 华而不实的

⑥ steady ['stɛdɪ] *adj.* 稳定的

⑦ rationing ['ræʃənɪŋ] *n.* 配给制

⑧ catchphrase ['kætʃfreɪz] *n.* 名言，流行口号

⑨ groundnut ['graʊndˌnʌt] *n.* 北美野豆

⑩ haunt [hɔːnt] *v.*（令人不愉快的事）萦绕在心头

⑪ catastrophe [kə'tæstrəfɪ] *n.* 灾难

One writer has described it as 'post-war Britain's equivalent[①] of the Millennium[②] Dome.' Another called it a 'complete fiasco[③]'. In 1945 Britain and its empire were in the grip of post-war austerity and suffering from severe food shortages. Most of all, the nation needed fat, and lots of it. From July that year, Clement Attlee's energetic new Labour administration held the reins[④] of power and its minister of food, John Strachey, helped cook up a grandiose[⑤] plan to solve the problem. Aimed at ensuring a steady[⑥] flow of cheap vegetable oil, he aimed to get the nation back on its feet. The answer, he believed, was nuts!

One of his first moves in office once he became minister of food in 1946 was to introduce bread rationing[⑦], which hadn't even been rationed in the war. 'Starve with Strachey' was a popular catchphrase[⑧] that year. So when the great groundnut[⑨] plan came along, he leapt on it. It was a scheme that was to haunt[⑩] him and his government for the rest of the decade and to end in rather comical catastrophe[⑪].

The original idea had come from Frank Samuel, managing director of the United Africa Company. Groundnuts, another name for peanuts, were an excellent source of oil and already

项目失败

填饱群众肚子的坚果计划

一位作家把这一计划称为“战后英国可与**千禧穹顶**相提并论的事件”。另一位作家称为“一败涂地”。1945年，英国陷入战后经济紧缩时期，食物严重短缺。最重要的是，国家需要脂肪，大量的脂肪。从那年7月起，以克莱门特·艾德礼（Clement Attlee）为首的充满活力的新工党内阁掌握了政权，粮食大臣约翰·斯特拉奇（John Strachey）协助出台了一项宏伟计划来解决粮食短缺问题。为了保证廉价植物油能够稳定流通，他要让国家重新站立起来。他相信，答案就是坚果！

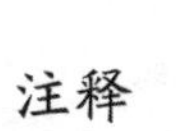

注释

千禧穹顶位于伦敦东部泰晤士河畔的格林尼威治半岛上，是英国政府为迎接21世纪而兴建的标志性建筑。

1946年，他上任粮食大臣后的第一步就是实行面包定量配给，在战争时期都没有这样过。“跟斯特拉奇一起挨饿”成为那年的流行标语。所以伟大的“落花生计划”一提出，他就开始实施。这项计划在之后10年缠住了他和他的政府，最终以相当滑稽的惨败收场。

这个想法最初是由非洲联合公司的总经理弗兰克·塞缪尔（Frank Samuel）提出来的。落花生（又名花生）是

grown on a small scale by locals in East Africa. But the region, most of which was still part of the British Empire, had vast tracts[①] of uncultivated land. All it needed, advised Samuel, was to plant this fast-growing crop and, using modern farming techniques, reap[②] the reward.

The resulting fats could be used to feed the British at home and bring wealth to Africa too. He thought it too big for a private firm to do and presented his findings to the government who commissioned a report from John Wakefield, former director of agriculture in Tanganyika, today's Tanzania. Wakefield's somewhat over-optimistic report recommended clearing 3.2 million acres for cultivation in the region. The sensible thing might have been to undertake a pilot scheme, as a junior minister suggested. But Strachey was in a hurry. Instead he gave the colossal[③] project a green light, authorising £25 million to cultivate 150,000 acres of scrubland[④] a year in Tanganyika, along with construction of a new port, railway and roads. The yield would be 600,000 tons of groundnuts annually, helping to boost the local economy and put a big dent[⑤] in Britain's food bill.

Almost nothing went right from the start. As Alan Wood, a journalist who wrote the definitive history of the affair back in the 1950s put it: 'they were proposing a colossal engineering and agricultural revolution, something comparable on a small scale to the Soviet Five-Year Plans, without even realising what they were doing.' No proper studies were undertaken on levels of rainfall in the area, essential for a crop that needed plenty of water. No thorough survey of the land or soil had been done or analysis of crop yields. These only started once the scheme was already well under way.

Work began in 1947 in the Kongwa region of the country. Yet there were not enough tractors to do the work, so the

注释

① tract [trækt] *n.* 大片
② reap [riːp] *v.* 获得
③ colossal [kəˈlɒsə] *adj.* 巨大的
④ scrubland [ˈskrʌbˌlænd] *n.* 灌木丛林地
⑤ dent [dɛnt] *n.* 凹痕

炼油的绝佳原料，而且东非当地人已经在小规模种植了。这一地区大部分仍然属于英帝国，还有广阔的荒地。塞缪尔建议种植这种快速生长的作物，并且采用现代农业技术收获结果。

产出的脂肪可以供英国人食用，也能给非洲带来财富。他认为这个项目对私有公司来说太大，就把他的发现递交给政府，政府委任坦噶尼喀（即现在的坦桑尼亚）前农业部部长约翰·维克菲尔德（John Wakefield）递交报告。维克菲尔德的报告有些过于乐观，建议开发这一地区的约130万公顷（320万英亩）空地。一位副部长建议先采取试点计划可能比较明智，但斯特拉奇迫不及待。他给这个庞大的项目大开绿灯，批准用2500万英镑在坦噶尼喀一年内开发约6万公顷（15万英亩）灌木丛林地，同时修建新的港口、铁路和公路。花生年产量将达到60万吨，有助于拉动当地经济，大幅消减英国的粮食开支。

整个计划从一开始就不太顺利。正如记者艾伦·伍德（Alan Wood）记载20世纪50年代发生的这件事的可靠历史时写道："他们提出了一场巨大的工程和农业革命，大约相当于小规模的苏联五年计划，而他们甚至都没意识到自己在做什么。"他们没有好好研究过那个地区的降雨量，这对需要充足水分的植物十分必要；没有全面勘测过那里的土壤；也没有分析过粮食产量。计划如火如荼实施之时，上述工作才刚刚展开。

工程于1947年在坦噶尼喀的孔瓜开工。然而拖拉机不够用，于是项目领导只好从世界各地拼凑二手拖拉机，包括美军在菲律宾多余的拖拉机。另一个问题是当地的铁

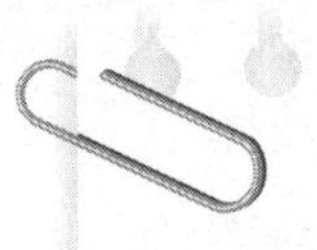

注释

① scour [skaʊə] *v.* 四处搜索
② surplus ['sɜːpləs] *n.* 过剩
③ transpire [træn'spaɪə] *v.*（人们）发现
④ envisage [ɪn'vɪzɪdʒ] *v.* 设想
⑤ bulldozer ['bʊlˌdəʊzə] *n.* 推土机
⑥ plague [pleɪg] *v.* 使困扰
⑦ scorpion ['skɔːpɪən] *n.* 蝎子
⑧ swarm [swɔːm] *n.* 大群（蜜蜂等昆虫）
⑨ hospitalize ['hɒspɪtəˌlaɪz] *v.* 送……住院治疗

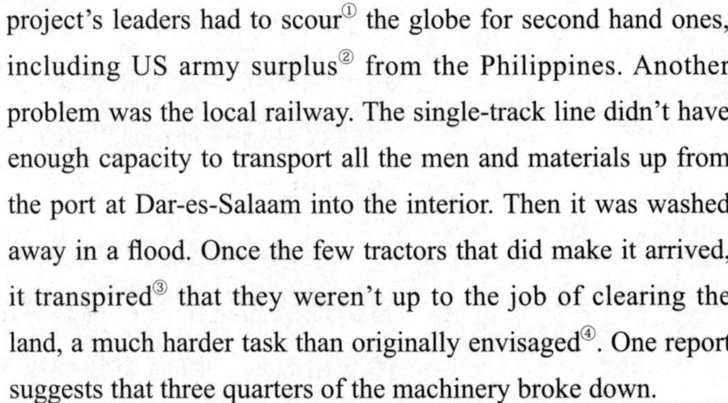

project's leaders had to scour[1] the globe for second hand ones, including US army surplus[2] from the Philippines. Another problem was the local railway. The single-track line didn't have enough capacity to transport all the men and materials up from the port at Dar-es-Salaam into the interior. Then it was washed away in a flood. Once the few tractors that did make it arrived, it transpired[3] that they weren't up to the job of clearing the land, a much harder task than originally envisaged[4]. One report suggests that three quarters of the machinery broke down.

When the team did find a way of clearing the trees that involved using three bulldozers[5] and a chain, they sent an order for ship's anchor chains to help. London cancelled the request thinking it was a joke. A vast army of thousands of men were engaged in the work. But they repeatedly went on strike. To top it all the workers were plagued[6] by scorpions[7] and swarms[8] of bees that left bulldozer operators hospitalised[9]. Agriculturally things weren't going much better. Once the groundnuts had been sown periods of drought turned the clay soil to concrete. By the end of the first year just 7,500 acres of groundnuts had been planted.

In spring 1948 new management was brought in under a new body, the Overseas Food Corporation, with a Major-General Desmond Harrison in charge on site. He tried to run the scheme as a military operation, but he soon had to return home sick. Towards the end of the project, which was finally abandoned in 1951, desperate officials tried growing sunflowers instead. There turned out to be too much sun and the crop failed. It was left to a new Minister of Food in a new government, Maurice Webb, to share the bad news about the scheme. One telling sentence in a government document of the time admitted: 'The groundnut is not a plant which lends itself readily to mass methods over vast acres.' It was also estimated that the crops had cost six times

路。单轨道铁路运力不足，无法把所有人员和物资从达累斯萨拉姆港运到内陆。后来铁轨又被洪水冲垮了。用为数不多的拖拉机完成运输任务之后，这时他们才发现无法完成清整土地的工作，这比原先设想的艰巨多了。一份报告表明，3/4的机械设备出现了故障。

项目组终于找到了一种清除树木的办法，需要用3辆推土机和1根链子，于是他们订购了船舶链。伦敦方面拒绝了这一请求，认为这简直是在开玩笑。成千上万的人力投入这项工作，但这些人不停地罢工。最糟糕的是工人们都受到了蝎子和蜂群的困扰，推土机驾驶员也因此住院了。农业方面情况进展也不好，花生刚种下，旱季就使得黏土变成了坚土块。到第一年年末为止，只种了约3035公顷（7500英亩）花生。

1948年春，新的管理团队来了，这是海外粮食公司，而且德斯蒙德·哈里森（Desmond Harrison）少将亲临现场坐镇指挥。他尝试以军队的管理模式来运作这一项目，但他很快不得不病怏怏地打道回府了。项目最终于1951年被放弃了，快要放弃前，绝望的官员们试图种植向日葵代替花生。最终由于日照过于强烈，种植向日葵的计划也失败了。这个项目留给了新内阁的新任粮食大臣毛里斯·韦勃（Maurice Webb），关于这个项目，他听到的也是坏消息。当时的一份政府文件中有句生动的话承认："花生是一种不愿意被大规模种植而用来满足大众的植物。"据估计，在种植花生过程中的花费是花生本身价值的6倍。

在随后的几年里，叫喊"花生"的次数多到足以让下议院发笑，而这个计划也成为政府重大工程失败的同义

more to produce than they were worth.

In the years that followed the cry of 'groundnuts' was enough to reduce the House of Commons to giggles[1] and the scheme became synonymous[2] with the failure of big government projects. Not so funny was the bill for the whole debacle[3], which came in it at around £49 million, estimated at more than £1 billion in today's money. Strachey's reputation never recovered.

For a government lauded[4] for achievements such as setting up the NHS, the great groundnuts cock-up is an episode the administration's fans would rather forget. Decades later Labour leader and thinker Michael Foot – a Labour MP in the late 1940s – bemoaned[5] the scheme's failure saying it could have been a blueprint for feeding Africa. He said: 'I think they could have gone ahead with many other schemes of that nature – both from the point of view of trying to assist the countries in Africa as well as to help the food policy here. Because that scheme failed they rather got cold feet, but I think that was a pity.'

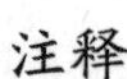

注释

① giggle ['gɪgəl] *v.* 咯咯地笑
② synonymous [sɪ'nɒnɪməs] *adj.* 密不可分的
③ debacle [deɪ'bɑ:kəl] *n.* 彻底失败
④ laud [lɔ:d] *v.* 嘉许，称赞
⑤ bemoan [bɪ'məʊn] *v.* 惋惜，埋怨

词。整个计划受损的账单就不那么好笑了，大约有4900万英镑，比现在的10亿英镑还多。斯特拉奇声名扫地，再无出头之日了。

对于一个因设立国民健康保险制度等成就而大受褒奖的政府来说，内阁的忠实追随者宁愿忘记花生工程搞得一塌糊涂这幕闹剧。几十年后，工党领袖及思想家迈克尔·富特（Michael Foot）——20世纪40年代的工党国会议员——为花生计划感到惋惜，说这个计划本可以成为解决非洲饥饿问题的蓝图。他说：“我认为，无论是为了帮助非洲国家，还是为了落实本国的粮食政策，他们本来可以继续实施其他类似性质的计划。那个计划失败了，他们就丧失了勇气和信心，我觉得十分可惜。”

ABANDONED

RAVENSCAR: THE HOLIDAY RESORT THAT NEVER WAS

The place 'has been happily named, offers a rare combination of nearly all the natural advantages required in a health resort, watering place, and holidaymakers[①]' home'. So read the brochure for a grand, new seaside town, one which aimed to trip off the tongue as easily as Blackpool or Brighton. It was to be another of the places to which smog-bound urban Victorians could easily flee. The florid[②] description of Ravenscar certainly made it seem like the ideal venue for a relaxing break away from the stresses and strains of city life. Yet there was something strange about the tempting[③] pleasures outlined in the brochure. For most of the luxurious resort of Ravenscar didn't yet exist.

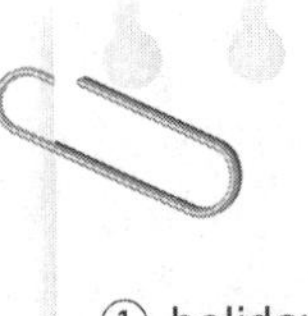

注释

① holidaymaker ['hɒlɪdeɪˌmeɪkə] *n.* 度假者
② florid ['flɒrɪd] *adj.* 过分花哨的
③ tempting ['tɛmptɪŋ] *adj.* 诱人的
④ promenade [ˌprɒmə'nɑːd] *n.* 散步场所
⑤ pier [pɪə] *n.* 凸式码头
⑥ bustle ['bʌsəl] *v.* 奔忙

The Victorians have been credited with inventing the idea of the seaside-holiday resort. Indeed most of the traditions associated with the British seaside such as the promenades[④], piers[⑤] and donkey rides emerged during the nineteenth century. Small towns and villages were transformed into bustling[⑥] popular destinations as people escaped the sooty cities created by the industrial revolution to enjoy the sea air and cheap entertainment. These new pleasure centres also became more

项目废弃

乌鸦崖：不存在的度假胜地

这个地方“有一个喜庆的名字，集中了疗养胜地、海滨浴场和度假疗养地所需的几乎全部自然优势，真是世所罕见”，一个新建海滨小镇的宣传册上如是写道。这个度假胜地希望拥有像黑水潭或布莱顿那样顺口的名字，希望成为维多利亚时代生活在雾霾中的城里人向往的另一处胜地。当然，这些华丽的描述让乌鸦崖看起来是一个摆脱城市生活压力、放松身心的好去处。然而，这本宣传册中那些诱人的描述听上去有些怪异，因为这个奢侈的乌鸦崖度假胜地的大部分纯属子虚乌有。

人们认为，维多利亚时代的人发明了海滨度假。的确，许多英国海滨度假传统，譬如海边散步、海滨游乐码头、沙滩骑驴等，都是19世纪出现的。那时，工业革命给城市带来了煤烟污染，人们开始逃离城市，去海边享受清新的空气和廉价的娱乐；小村镇也变得熙熙攘攘，受到人们的喜爱。同时，由于19世纪英国铁路的迅速发展，去这些新的娱乐热地度假更为便捷。

注释

① dissipate ['dɪsɪˌpeɪt] *v.* 驱散，消散
② hamlet ['hæmlɪt] *n.* 小村庄
③ stunning ['stʌnɪŋ] *adj.* 极美的
④ hilly ['hɪlɪ] *adj.* 多山丘的
⑤ castellated ['kæstɪˌleɪtɪd] *adj.* 类城堡的
⑥ pile [paɪl] *n.* 堆
⑦ tunnel ['tʌnəl] *n.* 隧道
⑧ omen ['əʊmən] *n.* 预兆
⑨ idyll ['ɪdəl] *n.* 田园生活，乡村生活

accessible thanks to the century's railway boom.

By the 1890s many resorts were well established and the appetite for the seaside showed no signs of dissipating[①]. The list of places which had cashed in on this boom was long. As well as Brighton and Blackpool was Skegness, Southport, Bridlington in the north and Bournemouth and Weymouth in the south – to name just a few. In the 1890s developers drew up plans for a new destination in Yorkshire designed to rival Scarborough, which had become another great seaside Mecca. The hamlet[②] of Peak, where the new town was to be constructed, was a stunning[③] location. Set 600ft up on a dramatic headland its location looked out over the beautiful Robin Hood Bay and behind it lay the wild expanses of the North Yorkshire Moors. Crucially, in 1885, the spot got a train station when the Scarborough and Whitby Railway was pushed through the hilly[④] landscape, making access for tourists straightforward.

The line had been built thanks to a William Hammond, who owned the local castellated[⑤] pile[⑥], Peak Hall. As chairman of the North Eastern Railways Company, he had been instrumental in the building of the new station near to his home. But perhaps the fact that he insisted on getting an extra tunnel[⑦] built so that he didn't have to see the line was an ill omen[⑧] for any future development of the rural idyll[⑨].

In 1895, following the death of Hammond, the couple's four daughters sold the hall and its land. The hall subsequently opened as the Raven Hall Hotel and a golf course was built. By 1897 The Peak Estate Company, led by an entrepreneur called John Bland, had bought the land for £10,000 and planned to make the hotel the hub of their new resort. A brickworks was opened nearby to supply the anticipated building boom. The idea was to sell 1,500 plots of land for houses and 300 men were soon put to work building sewers, mains water supply and

到19世纪90年代，英国已经出现了许多发展成熟的度假胜地，而人们对海滨度假的热情一如既往。靠这股度假热致富的地方多得不胜枚举，譬如，除了布莱顿和黑水潭，还有北部的斯凯格内斯、南港、布里德灵顿，以及南部的伯恩茅斯和韦茅斯。在90年代，开发商计划在约克郡建立度假点，与另一海滨旅游热镇斯卡布罗竞争。新的度假小镇准备驻扎在山峰村——一个美得令人窒息的小村落。它将被建在一个约183米（600英尺）高的引人注目的海岬上，那里俯瞰美丽的罗宾汉湾，背靠广阔的北约克湿地。最重要的是，在1885年，这个山区景点开通了斯卡布罗铁路和惠特比铁路，还建了火车站，游客可以更便捷地到达那里。

这条铁路线的修建，得益于一个叫威廉·哈蒙德（William Hammond）的人，他是当地城堡山峰庄园的主人。作为东北铁路公司的董事长，他曾协助修建了这座位于他家附近的新火车站。但是他坚持再修一条隧道，因为他不想看到铁路裸露在外，这对乡村田园的建设或许不是个好兆头。

1895年，哈蒙德去世，哈蒙德夫妇的四个女儿卖掉了庄园和土地。接着庄园被更名为乌鸦庄园酒店，对外营业。此外，那里还建了一个高尔夫球场。到1897年，在企业家约翰·布兰德（John Bland）的领导下，山峰房地产公司以1万英镑买下了这块地，准备把酒店改造成新的度假中心。为了给下一波建设热潮提供原材料，附近还成立了一间砖瓦厂。他们的计划是在这里出售1500块房屋土地。很快，300名工人就开始修建排水沟、自来水管，甚

even laying out a grid of streets. There was Roman Road, Saxon Road and some even got grand names like Marine Esplanade. As well as shops and tearooms there were to be hanging gardens too.

The name of the town was to be changed to Ravenscar after the Viking raiders[①] who had once plundered[②] the Yorkshire coast. The Vikings had used an emblem[③] of a raven[④] on their banners[⑤], while 'scar' was a Norse word for cliff. In a move that was certainly premature, you could even get a guidebook to the place. There appeared to be no question that Ravenscar would attract a multitude of visitors and investors eager to make the most of this spectacular addition to the nation's seaside attractions. The company's brochure continued: 'With the huge population of the West Riding and Midlands behind it, annually overcrowding the existing outlets to the sea, it needs little prescience to foresee that the future of Ravenscar as a watering place is practically assured.'

However, when prospective buyers for the properties arrived most were disappointed. What there was of a beach below the cliffs was rocky and Ravenscar's altitude meant that the shoreline was pretty inaccessible anyway. Despite the rather dubious[⑥] claim that in 1902 it had enjoyed seventy fewer rainy days than Scarborough, the climate could be very inhospitable. The cliffs were lashed by high winds and sea mists would often settle over the high ground. Prospective buyers travelling to inspect Ravenscar were offered a refund on their train fares if they bought a plot of land, but too few appreciated the charms of the place. In 1911, the company behind the scheme went bust.

Today the street layout is still evident. But where you'd expect to find rows of terraces[⑦] dotted with B&Bs, there are isolated Victorian houses which look as if they have been

注释

① raider ['reɪdə] *n.* 袭击者，侵入者

② plunder ['plʌndə] *v.* 掠夺

③ emblem ['ɛmbləm] *n.* 象征，标志

④ raven ['reɪv(ə)n] *n.* 渡鸦

⑤ banner ['bænə] *n.*（游行或集合用的）横幅

⑥ dubious ['dju:bɪəs] *adj.* 可疑的，不太可靠的

⑦ terrace ['tɛrəs] *n.*（互相连接的）排房屋，排屋

至布置交错的街道，有罗马路、撒克森路，一些名字还挺气派，像滨海大道。除了商店和小餐馆外，还会修建空中花园。

小镇的名字将被更名为乌鸦崖，以曾经在约克郡海岸劫掠财物的北欧海盗命名。这些北欧海盗用乌鸦（raven）作为他们的旗徽；而scar为古斯堪的纳维亚语，意思是峭壁。你甚至能得到这个地方的旅行指南，虽然现在就印刷旅行指南显然过早。看来人们相信乌鸦崖一定会吸引许多游客，还有那些想靠这个又一吸引眼球的海滨旅游胜地发财的投资者。公司的宣传册上写道："西部和中部地区人口众多，每年各个海滨旅游景点都会爆满，不难猜到未来的海滨浴场乌鸦崖肯定胜券在握。"

然而，当准备购买土地的投资者来到这里后，多数人都失望了。悬崖下面的海滩到处都是岩石；乌鸦崖又太高，游客很难到达海边。尽管据说在1902年此地雨天比斯卡布罗镇少70天（这种说法似乎有点靠不住），这里的气候条件却可能很恶劣。疾风会刮过峭壁，海雾也会时常光顾高地。公司许诺那些前来考察乌鸦崖的买主，如果他们买下一块地，就给他们报销火车票，但是没有几个人看得上这个地方。1911年，这家公司倒闭了。

今天，人们仍可以看到那里街道的布局。不过，本来应该是夹杂着民宿的齐整的排屋，今天却只有一些维多利亚式房屋，东一座，西一座，就像是从一些成功的度假胜地，如马盖特或莫克姆那里捡过来的。今天，乌鸦崖的人口只有几百人，它没能如人所愿成为一个热闹的小镇。由于理查德·毕钦（Richard Beeching）博士的报告带来的

注释

① pluck [plʌk] *v.* 摘，拔
② swingeing ['swɪndʒɪŋ] *adj.* 严厉的
③ boast [bəʊst] *v.* 吹嘘

plucked[①] from the streets of more successful resorts like Margate or Morecambe. The population of Ravenscar today is just a few hundred; it's not the bustling town that was hoped for. In 1965 even Ravenscar's station closed, a victim of the swingeing[②] Beeching cuts, based on the report by Dr Richard Beeching which saw many supposedly unprofitable local lines shut down. It was a far cry from the railway poster which had once proudly boasted[③] of the charms which awaited visitors to the Yorkshire paradise. It read: 'twixt moors and sea … magnificent undercliff and hanging gardens … most bracing health resort on the East Coast'.

“毕钦”大削减，许多被认为不赚钱的地方铁路线被停运。1965年，乌鸦崖的火车站也被关闭，完全不是铁路广告上曾吹嘘的等待游客到约克郡天堂旅游的情景了。那则广告上写着：“在湿地与大海之间……壮观的副崖和空中花园……东部海岸最令人心旷神怡的疗养胜地。”

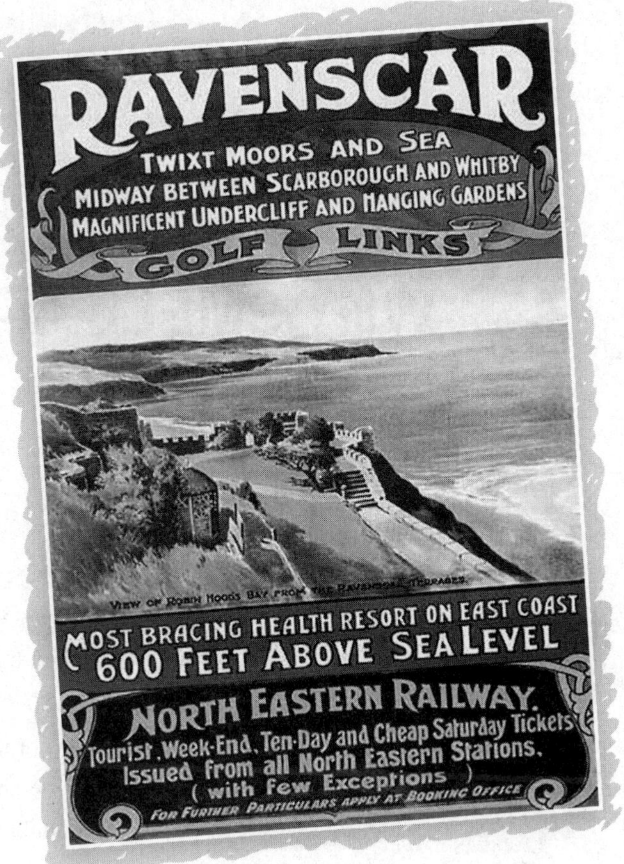

▲ 一份古旧的宣传乌鸦崖的铁路广告，乌鸦崖是一个不存在的约克郡海滨度假胜地。

A vintage railway poster advertising the charms of Ravenscar, the phantom Yorkshire seaside resort.

ABANDONED

WHY SMELL-O-VISION STANK

First there had been the movies, then there had been the talkies – now there would be the smellies! So went the hype for the much-anticipated 1960 movie *Scent of Mystery*. The film's producer Michael Todd Jr. believed that he was on the precipice of a new era in movie making, one in which audiences would not only be able to see the action but sniff① out the plot too. This would be achieved thanks to a revolutionary② new process which would enliven the action. Appropriate odours were to be pumped③ into the cinema as the movie's storyline developed.

The introduction of Smell-O-Vision, as this new technology was called, may now seem a potty④ episode⑤ in the history of cinema. But in the 1950s and 1960s studio bosses were worried about the growing lure⑥ of television and desperately trying to find ways to keep the cinemas full. Of the array⑦ of new technical ideas which aimed to enhance the big screen experience, Smell-O-Vision was the most intriguing⑧. The science behind it was provided by a Swiss inventor called Hans Laube. He'd already come up with a successful process of getting rid of smells in cinemas when it occurred to him that they could also be added back in too.

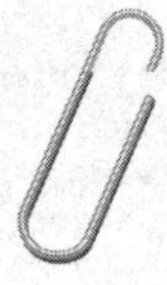

注释

① sniff [snɪf] *v.* 嗅
② revolutionary [ˌrɛvəˈluːʃənərɪ] *adj.* 革命的
③ pump [pʌmp] *v.* 抽送
④ potty [ˈpɒtɪ] *adj.* 愚蠢的，有点疯狂的
⑤ episode [ˈɛpɪˌsəʊd] *n.* 插曲
⑥ lure [lʊə] *n.* 诱饵，诱惑力，魅力
⑦ array [əˈreɪ] *n.* 大量，各种
⑧ intriguing [ɪnˈtriːgɪŋ] *adj.* 新奇的

项目废弃

为什么嗅觉电影惹人嫌

先是无声电影，再是有声电影，接下来是嗅觉电影！1960年备受期待的电影《神秘的气味》的广告是这样写的。电影制片人小迈克尔·托德（Michael Todd Jr.）相信，电影制作即将进入一个崭新的时代，观众不仅能够观看演员表演，还能用鼻子嗅出剧情。一种能使影片充满生气的革命性新技术将把这变成现实。随着影片剧情的展开，相应的气味就被送入影院。

在今天看来，这种称为嗅觉电影的新技术只是电影史上微不足道的一段小插曲。但是，在20世纪50—60年代，电影公司老板担心越来越流行的电视会取代电影，于是想尽一切办法维持票房。在一系列改进电影体验的新技术中，嗅觉电影是最有意思的。瑞士发明家汉斯·劳伯（Hans Laube）提出了这项技术。在成功发明了一种能去除电影院异味的技术后，他又想到也可以给电影院提供各种不同的气味。

他不是第一个尝试在电影放映厅刺激观众嗅觉的人。

注释

① olfactory [ɒl'fæktərɪ] *adj.* 嗅觉的

② auditorium [ˌɔːdɪ'tɔːrɪəm] *n.* 观众席

③ compliment ['kɒmplɪm(ə)nt] *v.* 恭维，称赞

④ spray [spreɪ] *v.* 喷（液体），（液体）喷出

⑤ lilac ['laɪlək] *n.* 丁香，丁香树

⑥ ventilation [ˌventɪ'leɪʃ(ə)n] *n.* 通风设备

⑦ smash [smæʃ] *adj.* 了不起的，非常轰动的，出色的 smash hit 轰动的演出，巨大的成功

⑧ proviso [prə'vaɪzəʊ] *n.* 限制性条款

⑨ catchy ['kætʃɪ] *adj.*（乐曲、名字、广告）引人注意的，容易记住的

⑩ fragrance ['freɪgrəns] *n.* 香味

⑪ trigger ['trɪgə] *v.* 引爆，触发

⑫ soundtrack ['saʊndˌtræk] *n.*（电影的）配音

He wasn't the first to try stimulating the olfactory[1] senses in movie auditoriums[2]. As far back as 1916, a newsreel about an American football game from LA's Rose Bowl had been complimented[3] by one cinema in Pennsylvania using rose oil and a fan. A Boston cinema sprayed[4] lilac[5] oil through its ventilation[6] systems over the opening credits of the 1929 film *Lilac Time*. Laube, however, was the first person to propose a precise system that delivered a whole range of odours to order. In 1939 he travelled to the New York Fair to demonstrate what he called his 'smell-brain'. He made a short film to show how smells could be delivered to individual cinema seats. The idea aroused some interest in the papers, but studio bosses remained unexcited.

Then, in the late 1950s, Todd Junior strode on to the stage. He was the son of famous movie mogul Mike Todd Senior and had worked with him on the wrap-around, big-screen phenomenon called Cinerama which used three projectors to produce a dramatic image. He'd also helped with his father's 1956 movie *Around The World In 80 Days*, which became a smash[7] hit and won an Oscar. After Todd Sr. died in a plane crash in 1958, Todd Jr. took up the reins of the business. He wanted to go further in the field of movie innovation. Having already discussed the idea of adding smells to films with his dad, Todd decided to employ Laube, paying for his experiments with a 'smellies' system at Chicago's Cinestage cinema. There was one proviso[8] – what Laube had been calling Scentovision was to be changed to the catchier[9] Smell-O-Vision.

Laube began refining his system. Smells were to be pumped to cinema-goers through tubes beneath their chairs using a series of vials filled with the different fragrances[10], which would be triggered[11] by the film's soundtrack[12]. Each vial contained 400cc of smell, enough for 180 performances. A mile of tubing in each

早在1916年，宾夕法尼亚州的一家电影院在播放一场在洛杉矶玫瑰碗举行的橄榄球比赛的新闻影片时，就使用了玫瑰油和风扇。一个波士顿电影院在1929年电影《丁香花季节》片头使用了通风设备喷洒丁香花油。不过，劳伯是提出包含一整套气味的精确体系的第一人。1939年，他前往纽约博览会展示他发明的“嗅觉大脑”。他制作了一个短片，向人们演示各种气味是如何被送至电影院观众席的。这个想法引起了一些报纸的兴趣，但制片厂老板却反应平淡。

20世纪50年代末，小托德登上了嗅觉电影的舞台。他是著名电影大亨老迈克·托德的儿子，他俩曾一起开发一种叫宽银幕立体电影的大屏幕环绕电影，这是一种使用三台同步放映机放映电影画面的技术。他也曾参与制作他父亲于1956年制作的电影《八十天环游地球》。这部影片大获成功，并赢得一项奥斯卡奖。1958年，老托德死于飞机失事，小托德掌管了整个公司。小托德想在电影创新上走得更远。那时他已经和父亲讨论过在电影中加入气味，他决定聘请劳伯，让他在芝加哥的影剧电影院进行他的气味体系试验，但有一个附加条件——劳伯的气味电影要改名为更好记的嗅觉电影。

劳伯于是开始改进他的气味系统。这个系统使用了一系列装有不同香味的小瓶子，在电影配乐的控制下，气味通过观众席下的管道被放出。每个小瓶里装有400厘米^{3}的气体，足够使用180次。每个剧院配有约1.6千米（1英里）长的气味管道，可将香气送入观众的鼻子。这个技术将在喜剧惊悚片《神秘的气味》中首次使用，这部电影也是按

theatre would get the aromas to the noses of the paying public. The process was to be debuted① as part of the comedy thriller *Scent of Mystery*, which was to be shot with Smell-O-Vision in mind. The film starred Denholm Elliott as a holidaymaker in Spain who uncovers a conspiracy② to murder an American heiress③, played by Elizabeth Taylor.

Thirty different odours would be emitted during the showing of the film. They ranged from garlic④ to tobacco and bananas as well as bread and shoe polish. When, in one scene, wine casks are smashed a grape fragrance was to be released. To add spice, the smells were worked into the plot too, providing clues to help Elliot's character get to the bottom of the mystery. Elizabeth Taylor, for example, is identified by her expensive perfume, the killer by the smell of his pipe. In the run up to the movie's release the press went wild for the idea. Todd helped whip up the frenzy⑤ with pun-filled publicity. 'I hope it's the kind of picture they call a scentsation!' he said. Once the film had debuted in three specially modified cinemas Todd had plans to speedily roll out the new technology to 100 more.

Heightening the tension, it transpired that Smell-O-Vision would be going head to head with a rival system – AromaRama. This had been introduced by Charles Weiss and Walter Reade Jr. for a documentary film *Behind The Great Wall* which hit cinemas in December 1959 just weeks before *Scent of Mystery* was released in February 1960. Weiss and Reade's smell mechanism delivered fifty-two odours⑥ such as jasmine⑦ into the cinema via the air conditioning system. The contest between the two was dubbed the 'battle of the smellies' by *Variety* magazine. In reality *Scent of Mystery* was a much bigger deal. Its system was more elaborate and cost three times as much to install. At its release in New York, Los Angles and Chicago, *Scent of Mystery* was even preceded by an animated short film

注释

① debut ['deɪbjuː] *v.* 初次登台，首次露面
② conspiracy [kən'spɪrəsɪ] *n.* 合谋
③ heiress ['ɛərɪs] *n.*（尤指大笔财产的）女继承人
④ garlic ['gɑːlɪk] *n.* 大蒜
⑤ frenzy ['frɛnzɪ] *n.* 疯狂
⑥ odour ['əʊdə] *n.* 独特气味
⑦ jasmine ['dʒæsmɪn] *n.* 茉莉

照嗅觉电影去拍摄的。影片讲述的是一个在西班牙度假的人——由丹霍姆·艾略特（Denholm Elliott）扮演，发现一个谋杀美国女继承人——由伊丽莎白·泰勒（Elizabeth Taylor）扮演的阴谋的故事。

在电影放映过程中，共要放出30种气味，从大蒜味到烟草味和香蕉味，还有面包味和鞋油味。有一幕是葡萄酒桶被击碎了，释放出了一股葡萄酒的香味。为了增添影片的趣味性，气味还被融入故事情节，帮助艾略特扮演的角色揭开故事谜底。譬如，伊丽莎白·泰勒身上有昂贵香水的气味，而杀手身上有烟斗的气味。在影片发行之际，媒体对这项新的电影技术进行了大肆宣传。托德用他的双关语给宣传又添了一把火。他说："我希望这部电影能'一炮打香'！"嗅觉电影在三个配有专门设备的影院放映后，托德计划在短时间内将这项新技术推广到100多家电影院中。

事实上，嗅觉电影还要与另一种气味电影体系——芳香电影——展开角逐，这让情势更加紧张。查尔斯·魏斯（Charles Weiss）和小沃尔特·里德（Walter Reade Jr.）在纪录片《中国长城》中使用了芳香电影技术，影片于1959年12月上映，比1960年2月上映的《神秘的气味》还早几个星期。魏斯和里德的技术是通过空调将52种像茉莉花那样的气味发送到影院。《多样性》杂志把这两种电影的竞争称为"气味之战"。事实上，《神秘的气味》的影响力更大，它使用的技术更复杂，安装成本也是《中国长城》的三倍。此外，《神秘的气味》在纽约、洛杉矶和芝加哥首映前，还播放了一个专门为嗅觉电影制作的动画短

注释

① loose-jointed [luːs ˈdʒɔɪntɪd] *adj.* 动作灵敏自如的，松松垮垮的
② disastrously [dɪˈzæstrəsli] *adv.* 悲惨地
③ hiss [hɪs] *v.* 发嘶嘶声
④ pervade [pɜːˈveɪd] *v.* 弥漫，充满
⑤ intermission [ˌɪntəˈmɪʃən] *n.* 幕间休息
⑥ furore [ˈfjʊərɔː] *n.* 狂怒，狂热
⑦ humdrum [ˈhʌmˌdrʌm] *adj.* 平凡的，单调乏味的，令人厌烦的
⑧ bemused [bɪˈmjuːzd] *adj.* 困惑不解的
⑨ baffle [ˈbæfəl] *v.* 使困惑

about a dog which had lost its sense of smell, also made for Smell-O-Vision.

The film's reviews were mixed. *The Hollywood Reporter*, for instance, announced that: 'the story itself has a distinct charm, fascination and humorous edge'. The *New York Times* critic Bosley Crowther was more harsh, branding it an 'artless, loose-jointed[①] "chase" picture set against some of the scenic beauties of Spain'. More disastrously[②], the Smell-O-vision process was not, as Todd had hoped, described as a 'scentsation'. Audiences complained of strange hissing[③] noises. Some were irritated that that the cinema's ventilation systems couldn't get rid of one smell before another was released. Others had difficulty picking up the odours which apparently led to the sound of heavy sniffing pervading[④] the cinema. *Time Magazine* summed it up like this: 'Most customers will probably agree that the smell they liked best was the one that they got during intermission[⑤]: fresh air.'

Laube hastily made adjustments, but it was too late. The *Scent of Mystery* wasn't making headlines at the box office and the furore[⑥] about the technology was fading. Some time later the movie was re-released with the much more humdrum[⑦] title *Holiday In Spain* and without the smells, which had the effect of leaving audiences even more bemused[⑧]. The *Daily Telegraph* reported: 'the film acquired a baffling[⑨], almost surreal quality, since there was no reason why, for example, a loaf of bread should be lifted from the oven and thrust into the camera for what seemed to be an unconscionably long time'.

In the coming years Todd gradually faded out of the movie business, moving to Ireland. Laube quietly disappeared from the scene, while the AromaRama system sank too. There were attempts in the ensuing decades to revive the idea. In the 1980s an Odorama version of the film *Polyester was* released with

片，讲述了一只失去嗅觉的狗的故事。

《神秘的气味》这部影片得到的评价毁誉参半。譬如，《好莱坞报道》评论说：“故事本身有其独特的情趣、魅力和幽默。”《纽约时报》评论家博斯利·克劳瑟（Bosley Crowther）则比较尖刻，将其形容为一部“以风景优美的西班牙为背景、毫无艺术性可言、情节松弛的‘追捕’电影”。更糟糕的是，嗅觉电影并没有像托德所希望的那样“一炮打香”。观众对放电影时奇怪的嘶嘶声感到不满。一些人抱怨电影院里的通风设备不好，前一种气味还没消除，后一种气味就出来了；由于另一些人闻不到气味，导致影院里到处都是鼻子抽搐的声音。《时代杂志》是这样总结这部电影的：“大部分观众或许会同意，他们最喜欢的气味是幕间休息时的气味——新鲜空气。”

劳伯仓促地做出一些调整，但已为时过晚。《神秘的气味》票房成绩不高，而这种技术的轰动效应也已逐渐消退。一段时间后，这部电影重新上映，片名改成了更乏味的《西班牙假日》，放映时也不再使用什么气味，但这次，观众感到更加困惑。据《每日电讯报》报道：“影片给人一种莫名其妙的、几乎是超现实主义的感觉。譬如，一块面包从烤箱里拿出来，然后镜头会毫无理由地长时间对着这块面包，这实在让人无法理解。”

之后几年，托德逐渐淡出电影圈，移居到爱尔兰，劳伯则销声匿迹，芳香电影从此一蹶不振。在此后的几十年里，也有人尝试再次让嗅觉电影进入影院。20世纪80年代，气味版电影《化纤品》上映，观众可以通过香味刮刮卡来闻各种气味。不过，迄今为止，《神秘的气味》仍是

scratch and sniff cards. To date, however, *The Scent of Mystery* remains the only movie to have been shot in Smell-O-Vision and it seems unlikely that the technology will be wafting① into multiplexes② anytime soon.

Japanese scientists have recently been working on a gadget③ that can accompany TV shows. But if a 1965 prank④ is anything to go by they might be wasting their time – the mere power of suggestion may be enough. That year the BBC played an April Fool's joke where it aired a spoof⑤ interview with a man who said he had invented a type of smell-o-vision for television, and then demonstrated by chopping some onions and brewing a pot of coffee. Duped viewers rang in to say that they had, indeed, smelled those very aromas⑥ coming through their TV sets.

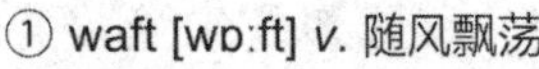

注释

① waft [wɒːft] *v.* 随风飘荡
② multiplex [ˈmʌltɪpleks] *n.* 多映厅影院
③ gadget [ˈgædʒɪt] *n.* 小器械（有时暗指复杂、不必要的东西）
④ prank [præŋk] *n.* 恶作剧
⑤ spoof [spuːf] *n.*（关于严肃事件的）玩笑性文章，玩笑性电视节目
⑥ aroma [əˈrəʊmə] *n.* 芳香，香味，香气

唯一一部专门为嗅觉电影摄制的影片，而短期内多厅电影院似乎不太可能飘出香味。

最近，日本科学家在研制一种气味装置，可配套电视节目使用。但是如果1965年的一个恶作剧可以给我们一些启示的话，那么这些科学家也许就是在浪费时间，因为通过暗示足以达到目的。那年BBC在愚人节给公众开了个玩笑，制作了一个假的访谈节目。受访者说他发明了一种嗅觉电视，然后当场演示切洋葱和煮咖啡。结果真有上当的观众打来电话说，他们通过电视闻到了洋葱和咖啡的气味。

BANNED
THE METAL CRICKET BAT

The soothing[①] sound of leather on willow. It is one of the joys of the sporting summer. The sound of leather on metal? It doesn't, somehow, have the same ring to it. Fiery Australian cricketing legend Dennis Lillee thought differently. When he walked out to bat at a sun-drenched[②] test match between England and Australia on 15 December 1979, he was wielding[③] a revolutionary blade[④] which would lead to one of the biggest controversies and on-pitch bust ups the sport had seen.

Lillee, one of the game's brightest stars and tempestuous[⑤] characters, had taken to the field at Perth's WACA stadium clutching[⑥] a bat made out of aluminium[⑦]. At the start of its second day, the match was delicately[⑧] poised[⑨] with Australia in some trouble on 232 for eight. Lillee was a superb fast bowler[⑩] who took 355 test wickets during his career. He was not, however, known as a great batsman. So as he strode to the middle with his new metal wand, perhaps he hoped that this bit of kit would add a sprinkle[⑪] of magic to his performance.

Lillee had ended the first day's play on a score of eleven not out, using an ordinary wooden bat. But on that second morning, he had chosen to take guard once more with the new

注释

① soothing ['su:ðɪŋ] *adj.* 抚慰的，使人宽心的
② sun-drenched ['sʌndrentʃt] *adj.* 阳光充足的
③ wield [wi:ld] *v.* 拿着（武器、工具或设备）
④ blade [bleɪd] *n.* 刃
⑤ tempestuous [tɛm'pɛstjʊəs] *adj.* 激烈的，狂暴的
⑥ clutch [klʌtʃ] *v.*（因为害怕或焦虑而）抓牢
⑦ aluminium [ˌæljʊ'mɪnɪəm] *n.* 铝
⑧ delicate ['dɛlɪkɪt] *adj.* 微妙的
⑨ poise [pɔɪz] *v.*（使）平衡，（使）悬着
⑩ bowler ['baʊlə] *n.* 投球手
⑪ sprinkle ['sprɪŋkəl] *n.* 撒，洒，少量

金属板球拍

皮革与柳木相撞，发出悦耳的声音。这是板球这种夏季运动给人们带来的乐趣。皮革与金属相撞呢？声音却没那么动听。性情暴躁的澳大利亚板球传奇人物丹尼斯·黎里（Dennis Lillee）却不这么认为。1979年12月15日，阳光明媚，英格兰队与澳大利亚队之间正在进行板球对抗赛，而黎里将在比赛中使用一种新型球拍，它将引起板球史上最大的争议和赛场上最严重的争吵。

黎里这位板球球坛上最闪耀的明星和脾气暴躁的人物，手拿铝制球拍走进了西澳板球协会珀斯球场。这是比赛第二天，澳大利亚以8:232的得分落后于英格兰。黎里是一个极快的投球手，他在他板球职业生涯中共有355次淘汰对手。不过，他还算不上一个出色的击球手。当他拿着他的新金属球拍大步走进球场时，他或许期待这个小装备能给他的表现增加一点儿魔力。

黎里在第一天比赛中得了11分未出局，用的是一根普通木球拍。但在第二天早上，他再次站在击球位上时，

invention. The ComBat, as it was called, had been created with the help of a businessman friend as a cheap alternative to the traditional wooden bats. Lillee had actually used the bat before in a test match, unleashing[①] it at a game against the West Indies in Brisbane, where it had only raised some wry smiles. It certainly didn't make much of an impact on the contest. The ball had clunked[②] against it only once, then thudded[③] against his pad leaving him out, leg before wicket[④], for no runs. Now, just two weeks later, an unperturbed[⑤] Lillee stood facing the equally charismatic and dangerous England bowler Ian Botham with the metal bat in his hands once again. On the fourth ball he received, Lillee saw his chance for a big shot, driving the ball towards the boundary, with England fielder David Gower in hot pursuit. As the ball left the bat, it made a distinctive, somewhat tinny sound.

England's captain Mike Brearley soon guessed what was up. Moments later he ran over to the umpires[⑥] Max O'Connell and Don Weser complaining that Lillee's bat was damaging the ball. A heated discussion ensued[⑦] with the umpires telling Lillee that he had to change the bat. Lillee stood his ground. Meanwhile, even Lillee's own captain, Greg Chappell, wasn't sure the bat was actually helping the Aussie cause. He thought the shot Lillee had hit, which went for three runs, would have gone for four if it had been hit with a normal willow[⑧] bat and promptly[⑨] sent his twelfth man, Rodney Hogg, out on to the pitch to deliver one to Lillee and retrieve the offending metal one. Still Lillee refused to budge[⑩].

For ten minutes debate raged in front of a bemused crowd, before Chappell himself intervened[⑪], marching on to the pitch with a wooden bat and ordering his team mate to use it. With his own skipper now against him, Lillee was persuaded to give in, but not before he'd hurled[⑫] his metal bat high into the air in

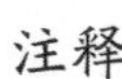

注释

① unleash [ʌn'li:ʃ] *v.* 释放
② clunk [klʌŋk] *v.*（重物相击时发出的）沉闷声
③ thud [θʌd] *v.* 发出沉闷声
④ wicket ['wɪkɪt] *n.*（板球中击球手）出球的一轮
⑤ unperturbed [ˌʌnpə'tɜ:bd] *adj.* 不受干扰的
⑥ umpire ['ʌmpaɪə] *n.*（体育比赛或竞赛的）裁判
⑦ ensue [ɪn'sju:] *v.* 随即发生
⑧ willow ['wɪləʊ] *n.* 柳树
⑨ promptly ['prɒmptlɪ] *adv.* 立即
⑩ budge [bʌdʒ] *v.* 使让步，做让步
⑪ intervene [ˌɪntə'vi:n] *v.* 干预
⑫ hurl [hɜ:l] *v.* 用力掷

用上了他的新发明。这个被称为康霸的新型板球拍是在他一个商人朋友的帮助下制作的，成本低廉，用于代替传统木制板球拍。事实上，黎里在以往的板球对抗赛中就用过这种球拍。在布里斯班与西印度群岛板球队比赛的时候，他的新型球拍曾亮过相，当时只引起了一些人的嘲笑。当然，它并没对比赛造成多大影响。这个金属球拍只有一次击中球，但是球随即砸到他的护垫上，腿碰球出局，没有得分。两周以后，此刻黎里镇定自若地站在同样魅力十足却极其危险的英格兰投球手伊恩·博瑟姆（Ian Botham）面前，手中再次握着那把金属球拍。当接第四个球的时候，他看到机会来了，于是把球打到边界，英格兰队外场手大卫·高尔（David Gower）紧随其后。当球离开球拍时，发出一声清脆、尖细的声音。

英格兰队队长迈克·布里尔利（Mike Brearley）很快就猜出是什么原因了。过了一会儿，他跑到裁判麦克斯·奥康纳（Max O’Connell）和唐·维瑟（Don Weser）那里，向他们投诉黎里的球拍会损坏板球。经过一番激烈讨论，裁判通知黎里，让他换掉球拍。黎里拒不让步。与此同时，黎里的队长格雷格·查普尔（Greg Chappell）也怀疑他的球拍是否真的能帮助澳大利亚队。他认为黎里那一击虽然得了3跑，但如果用普通柳木球拍他应该能得4跑。于是他让第12个队员罗德尼·霍格（Rodney Hogg）到球场给黎里送去一支普通球拍，取回那支违规的金属球拍。可黎里丝毫不让步。

激烈的争执持续了10分钟之久，而观众根本不知道发生了什么，直到查普尔本人插手。他拿着一支木制球拍走

disgust, 40 yards back in the direction of the dressing room.

Lillee went on to make eighteen runs, caught out off Botham's bowling, with the side all out for 244. Ironically Australia went on to win the game by 138 runs and won the three match series 3-0. Lillee may have had good reason to be disgruntled①. Using an aluminium bat wasn't actually against the game's rules. Law 6 stipulates② that the bat is 38in in length, and no more than 4¼ in wide. At the time it said nothing about what the bat was made from. Lillee might have pointed out that after equipment used in the world's major sports has always evolved. The first cricket bats were shaped like hockey sticks. Footballs started off as animal bladders③.

The spat over Lillee's bat actually turned out to be something of a brief marketing triumph, stoking not only controversy but sales of metal bats too. At the end of the match, Lillee had got all the players to sign the bat in question. Brearley wryly wrote on it: 'Good luck with the sales'. Lillee has since admitted that the ComBats were based on aluminium baseball bats and really designed: 'not for Test cricket but for practice, for clubs, for schools, all that sort of stuff '. Cricket traditionalists took a dim④ view of the whole incident. The game was only just recovering from the turmoil⑤ over Kerry Packer's controversial World Series Cricket. The row over the breakaway competition had briefly threatened to split the whole sport apart when Packer created a competition just for his Australian TV channels and lured players away from their national cricket associations.

Wisden Cricketers' Almanack, that guardian of the game's morals, said that the aluminium bat 'incident served only to blacken Lillee's reputation and damage the image of the game as well as, eventually, the Australian authorities because of their reluctance to take effective disciplinary⑥ action'. But the international game's unamused⑦ chiefs did take action. In 1980

注释

① disgruntled [dɪs'grʌntəld] *adj.* 生气的，不满的
② stipulate ['stɪpjʊˌleɪt] *v.* 规定，明确要求
③ bladder ['blædə] *n.* 膀胱
④ dim [dɪm] *adj.* 昏暗的
⑤ turmoil ['tɜːmɔɪl] *n.* 混乱，骚乱
⑥ disciplinary ['dɪsɪˌplɪnərɪ] *adj.* 纪律性的
⑦ unamused [ˌʌnə'mjuːzd] *adj.* 未被娱乐的，不笑的

进球场，命令球员使用这支球拍。黎里看到连自己的队长都反对他，只得让步，但他还是在离更衣室约40米（40码）远的地方，气愤地把球拍往天上使劲一扔。

黎里继续赢得18跑，然后被博瑟姆的投球接杀，澳大利亚队全部出局时得分为244分。具有讽刺意义的是，澳大利亚继续以138跑的成绩赢得比赛，最后以3比0获胜三场比赛。黎里则完全有理由对裁判的决定不满，使用铝制球拍并没有违反比赛规则。板球规则第六条规定球拍长不能超过96.52厘米（38英寸），球板宽不能超过10.8厘米（4.25英寸）。那时对制作板球拍的材料没有特别规定。黎里可以在事后提出，世界上大的体育项目使用的器材总是在不断改进。比如，第一个板球球拍的形状像曲棍球棒，足球最早的外形则像动物膀胱。

事实上，对黎里的金属球拍的争吵促成了短时期的商业成功——它不仅激起争论，还增加了金属板球拍的销量。比赛结束时，黎里让所有球员在他那个球拍上签名。球员布里尔利挖苦地在上面写道："愿它能卖得好。"从此以后，黎里承认康霸板球拍是根据铝制棒球球棒设计的，其用途"不是用于板球对抗赛，而是在练习中使用，或是用于俱乐部、学校等场合"。板球传统主义者对整个事件持不赞同意见。那时这项运动刚刚从凯瑞·帕克（Kerry Packer）有争议的世界板球系列赛的骚乱中恢复过来。帕克为他的澳大利亚电视频道专门设计了一种板球比赛，并吸引了许多来自全国板球协会的球员。而对于这个不同于传统板球比赛的新赛制的争论，已在短期内使整个板球运动面临分裂的危险。

注释

① debacle [deɪ'bɑːkəl] *n.* 彻底失败
② graphite ['græfaɪt] *n.* 石墨
③ strip [strɪp] *n.*（纸、金属、织物等）条，带
④ cringe [krɪndʒ] *v.* 感到局促不安

cricket's governing body changed Law 6 to state clearly that the bat should be made out of wood.

Metal bats do still exist, sometimes used in amateur forms of the game, but they are banned at test level. There were echoes of the Lillee debacle① in 2004, when another Australian, Ricky Ponting, started using a wooden bat that featured a graphite② strip③. This was subsequently banned too. Millions of cricketing traditionalists are thankful that Lillee's innovation never caught on. Nevertheless, the remaining ComBats are now sought after by collectors. Lillee has since suggested that throwing the bat in a fit of temper might have been a bit much. Shortly after releasing his autobiography, Menace, he said: 'I cringe④ a bit at it now.'

板球精神的守卫者《维斯登板球年鉴》这样评论道，铝制板球拍事件“只会给黎里的声望抹黑，还会损害板球运动的形象，最终损害澳大利亚官方形象，因为他们不愿意对此采取有效的纪律处分”。但是国际比赛中那些严肃的长官确实对此采取了行动。1980年，板球主管机构将第6条规则改为球拍应由木头制成。

金属板球拍现在仍然存在，有时用于业余比赛，但禁止用于对抗赛。2004年，也出现了类似黎里金属球拍的事情。另一个叫瑞奇·庞丁（Ricky Ponting）的澳大利亚人发明了一种有石墨条纹的木制球拍，后来也遭到禁止。千百万板球传统人士为黎里的创新没有流行起来而感到万分庆幸。然而，留存下来的康霸板球拍现在成了收藏家竞相收藏的对象。黎里后来也表示，自己当时发脾气把拍子一甩也许做得有点过火。在他的自传《威胁》出版后不久，他说道：“现在我对这件事感到有些羞愧。”

ABANDONED BICYCLE POLO & OTHER LOST OLYMPIC SPORTS

注释

① farce [fɑːs] *n.* 滑稽戏

② disqualify [dɪsˈkwɒlɪˌfaɪ] *v.* 使丧失资格

③ tournament [ˈtʊənəmənt] *n.* 锦标赛

④ tug [tʌg] *v.* 猛拉，拽

⑤ scoop [skuːp] *v.*（敏捷地）抱住

The summer Olympic Games of 1908 was a strange affair. Hastily held in London when Rome was forced to pull out for financial reasons following the eruption of Mount Vesuvius, it is probably best known for its marathon race which turned into something of a farce[①]. The winner, Dorando Pietri, collapsed as he entered the White City Stadium on the last leg of the race and was helped over the line by officials. The Italian runner was later disqualified[②].

In many ways the modern version of the Olympic Games, reintroduced in Athens in 1896, was still finding its feet. Both the 1900 and 1904 tournaments[③] had been held as part of World Fairs, rather than as events in themselves. And in those early days of the ultimate sporting spectacle some unusual disciplines were included in the programme. At the 1900 Games, held in Paris, the tug[④] of war was a big draw – in which, rather oddly, a mixture of Danes and Swedes teamed up against the French to scoop[⑤] the gold medal. The Games also hosted an array of 'demonstration sports' vying for their official inclusion which included cannon shooting, fire fighting and even ballooning.

In that year the French had also been the only country to take

自行车马球及其他消失的奥运项目

1908年的夏季**奥林匹克运动会**有些奇怪。由于维苏威火山爆发导致意大利财政困难，主办方罗马被迫放弃举办奥运会，转而在伦敦仓促举行。这次奥运会最有名的事件当属马拉松比赛了，这场比赛最后成了一场闹剧。第一名多兰多·皮特里（Dorando Pietri）在进入白城体育场快接近终点时倒下，被工作人员搀扶着穿过终点线。这位意大利选手后来被取消了比赛成绩。

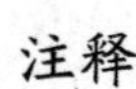

注释

奥林匹克运动会简称“奥运会”，是国际奥林匹克委员会主办的世界规模最大的综合性运动会，每四年一届。奥林匹克运动会发源于古希腊，因举办地在奥林匹亚而得名。

自从1896年奥运会圣火在雅典再次点燃之后，现代奥林匹克运动会在许多方面仍在摸索经验。1900年和1904年的比赛与其说是奥运会，不如说是世界博览会的一部分。在这个最高体育盛会的早期，出现了一些奇特的比赛项目。1900年巴黎奥运会上，拔河比赛非常引人注目，比赛中出现了奇怪的一幕，丹麦人和瑞典人组成联队与法国人争夺金牌。这次奥运会还举办了一系列表演项目，包括火炮射击、消防比赛，甚

part in the official croquet[①] competition, taking all the medals. Four years later, the Games were being held on US soil, in St Louis, Missouri for the first time. Rather than try and beat the French at their ancient pastime, the Americans decided to invent their own version of croquet. They simply removed the letters c and t from the beginning and end of the word croquet to come up with roque[②]. Though they tinkered[③] with the rules a little, the crucial difference was that roque, unlike croquet, was played on a hard surface rather than grass. Only four players took part in the roque competition – all American. In the following years the sport briefly caught the public's imagination; by 1920 there was a national league. But as the century wore on interest gradually fizzled[④] out.

The 1908 Games in London would see more intrigue[⑤]. Its tug-of-war event almost turned into a riot[⑥] when the US squad[⑦] accused one of three British teams, made up of Liverpudlian policemen, of wearing spiked[⑧] boots. The protest was overturned and the three different British teams went on to win gold, silver and bronze. Among the new sports was running deer, a shooting event which used targets rather than actual animals. Another sport that featured in 1908 was bicycle polo. This brand new game was invented by an Irishman and a retired champion cyclist Richard J. Mecredy. In 1891 he came up with a new spin[⑨] on an old sport. 'Posh polo' had been around for centuries, though equestrian[⑩] polo's official rules had only been drawn up in 1874. Why not, thought Mecredy, combine the drama of polo with the popular passion for the bike and create a 'common man's version of the sport of kings'?

As editor of a magazine called the *Irish Cyclist*, Mecredy published rules for the game in October 1891, following the first ever match between two Irish teams: Rathclaren Rovers and Ohne Hast Cycling Club. Early photos show players armed

注释

① croquet ['krəʊkeɪ] *n.* 槌球游戏
② roque [rəʊk] *n.* 短柄槌球
③ tinker ['tɪŋkə] *v.* 小修改
④ fizzle ['fɪzəl] *v.* 虎头蛇尾地结束
⑤ intrigue [ɪn'triːg] *v.* 激起……的好奇心
⑥ riot ['raɪət] *n.* 暴乱
⑦ squad [skwɒd] *n.*（专门处理某类犯罪的）警察分队
⑧ spiked ['spaɪkt] *adj.* 有尖顶的
⑨ spin [spɪn] *n.* 旋转
⑩ equestrian [ɪ'kwɛstrɪən] *adj.* 骑马的

至还有热气球比赛。

那年，法国也是唯一一个参加正式槌球比赛项目的国家，并囊括了这个项目的所有奖牌。四年以后，奥运会首次在美国密苏里州圣路易斯市举办。美国人没有挑战法国人这项古老的休闲运动，而是决定发明本国的槌球运动。他们仅将单词槌球（croquet）的首尾字母“c”和“t”去掉，发明了自己的槌球（roque）。虽然只是稍微改了一下规则，但美式槌球的最大不同之处是在硬地上而不是草地上进行。当时只有四个人参加了美式槌球比赛，他们都是美国人。后来，这项运动很快就引起了公众的兴趣，到1920年，还成立一个国家联队。但是随着时间的推移，人们对这项运动的热情逐渐消退。

1908年的伦敦奥运会更是充满奇闻。这届奥运会的拔河比赛几乎演变成一场骚乱：美国队投诉三支由利物浦警察组成的小队穿钉靴上阵，他们的抗议遭到否决，而三支英国队继续比赛，分享了金、银、铜牌。在新的比赛项目中有不使用动物活靶的射击比赛。1908年奥运会的另一个特色是自行车马球项目。这项全新的运动是一个爱尔兰人、退役的自行车赛冠军理查德·J. 米克里第（Richard J. Mecredy）发明的。1891年，他对一种旧运动进行了创新。“上流社会的马球”已经盛行了几个世纪，而马术马球的官方规则直到1874年才制定。米克里第想，为什么不能将精彩的马球与流行的自行车运动结合起来，创造一种“王室运动的平民版本呢”？

米克里第是杂志《爱尔兰骑车者》的编辑。在两支爱尔兰球队——拉斯克拉恩流浪者队和奥那哈斯特自行车

注释

① headgear ['hɛdˌɡɪə] *n.* 帽子，头饰
② manoeuvre [mə'nuːvə] *v.*（熟练地）移动

with mallets but no protective headgear[①]. A contemporary report reassured the public that the game 'was not at all so dangerous as would appear from the title'. Over the course of the next decade bicycle polo caught on in Britain too, with clubs in Catford, Newcastle and Northampton and a national association formed. In 1901 the first international game was played in which the Irish soundly defeated England 10–5.

When the 1908 Games came along, bicycle polo's fans saw an opportunity to boost the sport's profile further, getting it accepted as a so-called demonstration sport, which allowed hopeful Olympic sports to promote their cause for regular inclusion. Ireland, which still dominated the world of bicycle polo, wasn't allowed to compete in most events that year as a separate nation as it was still part of the United Kingdom of Great Britain and Ireland. But with tensions over Irish independence running high, Ireland was allowed to field its own teams in hockey, polo, and in bicycle polo too. At the final of the bicycle polo at the Games in July 1908 an Irish team, which appears to have included Mecredy's own son, soundly beat a German side 3–1.

Bicycle polo did not feature at the fifth Olympiad in 1912 in Stockholm and, thanks to the First World War, the sport lost many of its finest players. However, by the 1930s, the popularity of bicycle polo was growing in Britain once again and regional leagues were introduced. By 1938 there were around 170 official teams across the country and 1,000 registered players. A guide from the time spells out the rules of how a match should be played. Two teams feature four mounted players on a marked-out field measuring 100 yards by 60 yards. Games, lasting ninety minutes, would see players in pursuit of the ball while skillfully manoeuvre[②] their two-wheeled steeds. Each player wielded a wooden mallet with a head of 7in, aiming to

俱乐部——举行第一场自行车马球比赛之后，他在1891年发表了这项运动的规则。早年照片显示，球员使用槌棒打球，但没戴保护头盔。当时还公布了一份报告，说这项运动“没有它的名字听上去那么可怕”，让公众放心。在接下来的10年，自行车马球在英国也流行起来。在卡特福德、纽卡斯尔和北安普顿成立了俱乐部，还成立了一个全国自行车马球协会。1901年，举行了第一场国际自行车马球比赛，结果是爱尔兰队以10:5大败英格兰队。

随着1908年奥运会的到来，自行车马球的球迷认为这是进一步推广这项运动的大好机会。在他们的努力下，自行车马球成为一个表演项目，并希望借此被纳入奥运会的正式比赛项目。爱尔兰那时仍旧是自行车马球运动强国，但由于它仍然是大不列颠及爱尔兰联合王国的一部分，所以没有资格作为一个独立国家参加大部分比赛项目。不过爱尔兰独立的呼声高涨，使得爱尔兰允许在曲棍球、马球和自行车马球项目方面有独立的球队。在1908年7月奥运会的自行车马球决赛中，爱尔兰队以3:1大败德国队，当时好像米克里第的儿子也是爱尔兰队的一员。

然而，在1912年于斯德哥尔摩举办的第五届奥运会上，自行车马球没有露脸，而且由于第一次世界大战，这项运动失去了许多优秀的球员。不过，到了20世纪30年代，自行车马球再次在英国流行起来，还出现了地区联赛。到1938年，全国就有约170支正式球队和1000名注册球员。那时还制定了指南，列出比赛规则：双方各派4名队员，在一个长约91米（100码）、宽约55米（60码）的场地进行比赛。比赛计时90分钟，双方球员熟练地操纵

knock the ball into the opposing goal. Foul play was definitely out – on no account should the mallet① be shoved through an opposing player's spokes.

This heyday② was short lived. Again war intervened and after the Second World War the sport faltered③ as a popular pastime, but just about clung④ on, especially in France. While it never again graced an Olympic stage, and it's pretty unusual to see a match being played at your local park today, an urban version of the game, played on hard courts, has found a new wave of popularity in recent years. There are still international matches, even for the grass version.

It is, of course, possible that one day bicycle polo might re-emerge at the Olympics. After all golf, not played at the Olympics since 1900, is to return for the 2016 Games in Rio de Janeiro. There is one lost Olympic event that almost certainly won't be making a comeback – live pigeon shooting. It featured once, in 1900. To date it remains the only Olympic event from the modern Games in which live animals were killed. Almost 300 were shot. A Belgian, Leon de Lunden, won gold at the gory⑤ competition, felling a total of twenty-one birds.

注释

① mallet ['mælɪt] *n.* 木槌，（槌球和马球运动的）球棍，球棒
② heyday ['heɪˌdeɪ] *n.* 全盛时期
③ falter ['fɔːltə] *v.* 衰退
④ clung [klʌŋ]（cling）的过去式和过去分词 cling [klɪŋ] *v.* 抓紧，紧握
⑤ gory ['gɔːrɪ] *adj.* 血淋淋的

他们的两轮坐骑追逐马球。每个球员挥舞一根杆头约2.1米（7英尺）长的木槌，目标是将马球打入对方球门。犯规者肯定会出局——绝不允许将球杆插入对方球员自行车的轮辐。自行车马球的全盛期非常短暂。战争再次爆发，第二次世界大战后这项运动退变为一种时尚的休闲运动，但仍有人想重振旗鼓，特别是在法国。虽然它没能再次荣登奥林匹克舞台，而且今天在普通人家附近的公园也很少看到这种比赛，但是近年来一种在硬地赛场进行的城市自行车马球运动开始流行起来。现在还有国际自行车马球比赛，甚至有的还可以在草地上进行。

当然，自行车马球仍有可能重返奥运赛场。毕竟，高尔夫球比赛将在2016年里约热内卢奥运会再次出现，而这项运动自1900年以来就再也没有在奥运会上出现过。有一项被取消的奥运会项目肯定不会再来——活鸽射击。这个项目只在1900年奥运会上出现过一次。迄今，这是现代奥运会唯一一项杀死活物的运动，几乎射杀了300只鸽子。比利时人莱昂·德·兰登（Leon de Lunden）在那场血淋淋的比赛中击毙了21只鸽子，赢得金牌。

ILLUSTRATIONS
插 图

We would like to express our thanks to everyone who has helped with the illustrations in this book. We have made every attempt to trace owners of copyright material and to attribute each correctly. We apologise for any omissions and will be pleased to incorporate missing acknowledgements in any future editions.

我们谨向为本书提供插图的所有人表示感谢。我们尽了最大努力，希望找到版权资料的所有者并予以适当感谢。对于存在的疏漏，我们深表歉意，但愿在今后的版本中补上遗漏的致谢。

编者注

考虑到读者依据原文检索更为方便，故对本书插图提供机构名称及人员姓名未作翻译。

London's Eiffel Tower 伦敦的埃菲尔铁塔

Images courtesy of Brent Archives, London Borough of Brent, www.brent.gov.uk

Nelson's pyramid 纳尔逊的金字塔

The Royal Collection © 2011 Her Majesty Queen Elizabeth II

The tumbling abbey habit 放山修道院的倒塌

John Britton, 'Fonthill Abbey from South West', Graphical and Literary Illustrations of Fonthill Abbey (Wiltshire, 1823), courtesy of Beckford Tower Trust

Why Lutyens' cathedral vanished 为什么鲁琴斯的大教堂消失了

Images courtesy of National Museums Liverpool, www.liverpoolmuseums.org.uk

New York's doomed dome 注定失败的纽约穹顶

Image courtesy of The Estate of R. Buckminster Fuller

The brand new continent of Atlantropa 亚特兰特洛帕新大陆

Image courtesy of Popular Science magazine, The Bonnier Corporation

The perpetual motion machine 永动机

Image courtesy of the University of Kentucky Archives

Bentham's all-seeing panopticon 边沁的全视圆形监狱

Image copyright Harry Gates

Tesla's earthquake machine 特斯拉的地震机

Original patent image from the US Patent Office

Edison's concrete furniture 爱迪生的水泥家具

US Department of Interior, National Park Service, Thomas Edison National Historical Park

The X-ray shoe-fitting machine X射线试鞋机

Images courtesy of the National Museum of Health and Medicine, Washington D.C., http://nmhm.washingtondc.museum

Escape coffins for the mistakenly interred 避免误埋的逃生棺材

Original patent image from US Patent Office

Chadwick's miasma-terminating towers 查德威克的臭气终结塔

Jon Heal, Chartered Institute of Environmental Health

Ravenscar: the holiday resort that never was 乌鸦崖：不存在的度假胜地

Contemporary railway poster from the 1900s

SOURCES AND RESOURCES

资料来源

London's Eiffel Tower　伦敦的埃菲尔铁塔

Original publicity materials in the Brent Borough Council Archive.

Country Life (19 May 1955).

Daily Graphic (14 April 1894).

The Guardian (14 March 2006).

The first Channel Tunnel　第一条英吉利海峡隧道

Hansard (HL Deb 24 January 1929, Vol. 72), pp. 788–95.

Contemporary Review (1 June 1995).

The Independent (6 May 1994).

History Today (1 November 1996).

New Scientist (4 February 1971).

The Railway News (13 December 1873, 23 April 1883).

Telegraph (6 August 2010).

The Independent (2 April 2007).

编者注

考虑到读者依据原文检索更为方便，故对本书资料来源未作翻译，但依据《信息与文献参考文献著录规则》对原书文献著录格式进行了修改，并在能检索到的范围内对原书文献中的缺项信息进行了增补。

Nelson's pyramid 纳尔逊的金字塔

Felix Barker and Ralph Hyde, *London as it might have been* (John Murray, 1982).
G.H. Gator and F.R. Holmes, (eds), *Survey of London* (Vol. 20, 1940), on British History online, www.british-history.ac.uk
C.L. Falkiner, '(Richard) William Steuart Trench 1808–1872', *Dictionary of National Biography*.
Roy Porter, *London: a social history* (Penguin, 2000).
'Minutes of the evidence of the select committee on Trafalgar Square, 1840'.

Wren's missing marvels 雷恩错失的奇迹

Felix Barker and Ralph Hyde, *London as it might have been* (John Murray, 1982).
Kerry Downes, 'Christopher Wren 1632–1723', *Dictionary of National Biography*.
Antonia Fraser, *King Charles II* (Weidenfeld & Nicholson, 1989).
Stephen Inwood, *A History of London* (Macmillan, 1998).
Adrian Tinniswood, *His invention so fertile: a life of Christopher Wren* (Jonathan Cape, 2001).

The tumbling abbey habit 放山修道院的倒塌

John Rutter, *Delineations of Fonthill and its Abbey* (self-published 1823, facsimile published by Gregg Publishing, 1973).
Henry Venn Lansdown, 'Recollections of William Beckford: on Beckfordiana – a contemporary assessment by a friend of Beckford's', on the William Beckford Society website, www.beckford.
c18.net
William Donaldson, *Brewer's Rogues, Villains and Eccentrics* (Weindenfeld & Nicholson, 2002).
E.A. Smith, *George IV* (Yale, 1999).

Why Lutyens' cathedral vanished 为什么鲁琴斯的大教堂消失了

Liverpool Museums Publicity documents 2007.
Liverpool Metropolitan Cathedral website, www.liverpoolmetrocathedral.org.uk
Time Magazine (12 August 1929).
The Independent (9 July 1966).
Liverpool Daily Post (28 November 2006).
Apollo (1 January 2007).

New York's doomed dome 注定失败的纽约穹顶

R.M. Marks, *The Dymaxion World of Buckminster Fuller* (Southern Illinois University Press, 1960).
H. Kenner, *Bucky: A guided tour of Buckminster Fuller* (William Morrow &Co., 1973).
The Buckminster Fuller Institute at Santa Barbara, www.buckminsterfuller.net

The brand new continent of Atlantropa 亚特兰特洛帕新大陆

Willy Ley, *Engineers Dreams* (1955).
Cabinet Magazine (Spring, 2003).
Gibraltar Magazine (January 2009).
The Independent (24 July 2004).
Star-Tribune (14 July 1999).

A nation built on sand 建在沙上的国家

The Economist (24 December 2005).
Harper's Magazine, (October 2004).
Cabinet Magazine (Issue 18, 2005).

Paxton's orbital shopping mall 帕克斯顿的轨道购物商场

Independent (5 May 2009).
Telegraph (29 June 2005).
Telegraph (12 August 2003).
Georg Kohlmaier, Barna von Sartory and John C. Harvey, *Houses of Glass: A Nineteenth Century Building Type* (The MIT Press, 1991).
The Civil Engineer and Architect's Journal (Vol. 18, 1855).
James Winter, *London's Teeming Streets* (Routledge, 1993).
'Minutes of Evidence from the Select Committee on Metropolitan Communications, 1855.'

The perpetual motion machine 永动机

William H. Huffman, *Robert Fludd and the end of the Renaissance* (Routledge, 1988).

Arthur W.J.G. Ord-Hume, *Perpetual Motion: The History of an Obsession* (Adventures Unlimited Press, 1977).
Michael D. Lemonick, 'Will Someone Build A Perpetual Motion Machine?', *Time* (10 April 2000).

Bentham's all-seeing panopticon 边沁的全视圆形监狱

London's Bentham Project, University College: www.ucl.ac.uk/Bentham-Project
Internet Encyclopedia of Philosophy: www.iep.utm.edu/bentham
Simon Schama, *A history of Britain: Volume 3, The fate of Empire 1776–2000* (BBC, 2002).
A.N. Wilson, *The Victorians* (Arrow Books, 2003).

Tesla's earthquake machine 特斯拉的地震机

Margaret Cheney, *Tesla: Man Out of Time* (Simon & Schuster, 2001).
David Hatcher Childress, (ed.), *Anti-gravity handbook* (Adventures Unlimited Press, 1994).
Earl Sparling, 'Nikola Tesla, At 79, Uses Earth To Transmit Signals: Expects To Have $100,000,000 Within Two Years', *New York World Telegram* (11 July 1935).
Nikola Tesla Museum, Belgrade, holds a number of Tesla's documents.

Edison's concrete furniture 爱迪生的水泥家具

R.W. Clark, *Edison: the man who made the future* (New York, 1977).
'Edison now making concrete furniture', *New York Times* (9 December 1911).
Michael Peterson, 'Thomas Edison's Concrete Houses', *American Heritage* (Vol. 11, Issue 3, winter 1996).

The X-ray shoe-fitting machine X射线试鞋机

Leon Lewis and Paul E. Caplan, 'The shoe-fitting fluoroscope as a radiation hazard', *California Medicine* (Vol. 72, No 1, January 1950).
Council A. Nedd II, 'When the Solution Is the Problem: A Brief History of the Shoe Fluoroscope', *American Journal of Roentgenology* (June 1992).
Charles Hartridge, (letter by), 'Shoe X-Ray machines', *New Scientist* (3 March 1960).
'Fluroscope a success', *New York Times* (12 May 1896).
Dr Michael J. Smullen and Dr David E. Bertler, 'Basal cell carcinoma of the sole: possible association with the shoe-fitting fluroscope', *Wisconsin Medical Journal* (2007).

The cure that killed　致命的疗法

Paul W. Frame, *Radioactive curative devices and spa* (Oak Ridge Associated Universities, 1989).
Dr Mark Neuzil and Bill Kovarik, *Mass Media & Environmental Conflict* (Sage Publications, 1996).
Laura Sternick, 'Liquid Sunshine: The Discovery of Radium', *Journal of Science* (2008).
Nanny Fröman, 'Marie and Pierre Curie and the Discovery of Polonium and Radium', on www.nobelprize.org
'Health Consultation: US Radium Corporation', State of New Jersey (June 1997).
'Poison Paintbrush', *Time Magazine* (4 June 1928).
'Radium Drinks', *Time Magazine* (11 April 1932).
'Make luminous drinks from radium', *New York Times* (14 January 1904).

The 'cloudbuster'　破云器

Mildred Edie Brady, 'The Strange Case Of Wilhelm Reich', *Bulletin Of The Menninger Clinic* (March 1948).
R.Z. Sheppard, 'A family Affair: review of *A book of dreams* by Peter Reich', *Time* (14 May 1973).
Wilheim Reich obituary, *Time* (18 November 1957).
Judge Clifford, 'Decree Of Injunction Order' (USA vs. Wilhelm Reich) (March 19, 1954).
Wilheim Reich, *Die Bione* (1938). English translation on: www.actlab.utexas.edu/~hahmad/bions.htm

Escape coffins for the mistakenly interred　避免误埋的逃生棺材

Jan Bondson, *Buried Alive: The Terrifying History of Our Most Primal Fear* (W.W. Norton, 2002).
Bill Bryson, *At Home* (Doubleday, 2010).
Stephen B. Harris, 'The Society for the recovery of persons apparently dead', *Cryonics* (1990) (at the Alcor Life Extension Foundation).

Chadwick's miasma-terminating towers　查德威克的臭气终结塔

Stephen Halliday, 'Death and miasma in Victorian London, an obstinate belief ', *British Medical Journal* (22 December 2001).
Di Samuel Edward Finer, *The life and times of Sir Edwin Chadwick* (Methuen, 1952).
Vladimir Jankovic and Michael Hebbert, 'Hidden Climate Change: Urban Meteorology and the Scales of Real Weather', Paper delivered at the Royal Geographic Society-IRG annual conference, 2009.
Horace Joules, 'A preventative approach to common diseases of the lung', *British Medical Journal* (27 November 1954).
A.N. Wilson, *The Victorians* (Arrow Books, 2003).

The self-cleaning house 自洁式房屋

Colin Davies, *The Prefabricated Home* (Reaktion Books, 2005).
Kristin McMurran, 'Frances Gabe's Self-Cleaning House could mean new rights of spring for housewives', *People Magazine* (29 March 1982).
Kim Elliott, 'Frances Gabe: At home in Futureville' (2009), on www. serendipityjones.com
'Vatican paper: washing machine liberated women most', *Reuters* (9 March 2009).

The jaw-dropping diet 令人目瞪口呆的饮食法

Horace Fletcher, *The A.B.Z of Our Own Nutrition* (1903).
Horace Fletcher, *Fletcherism: What is it or how I became young at Sixty* (1913).
David F. Smith, (ed.), *Nutrition in Britain* (1997).
Daily Mail (14 March 2008).
New Scientist (6 November 1980).
New York Times (14 July 1919).

A nutty plan to feed the masses 填饱群众肚子的坚果计划

History Today (1 January 2001).
Alan Wood, *The Groundnut Affair* (Bodley Head, 1950).
D.R. Myddelton, 'They Mean Well: Government Project Disasters', *The Institute of Economic Affairs* (2007).
The Independent (15 January 2010).

Ravenscar: the holiday resort that never was 乌鸦崖：不存在的度假胜地

The Northern Echo (19 February 2001).
Yorkshire Post (13 November 2004).

Why Smell-O-Vision stank 为什么嗅觉电影惹人嫌

Mark Thomas, *Beyond Ballyhoo: Motion Picture Promotion and Gimmicks* (McGee, McFarland, 2001).
Scott Kirsner, *Inventing the Movies* (CinemaTech Books, 2008).
The Independent (10 May 2002).
New York Times (8 May 2002).

Martin J. Smith and Patrick J. Kiger, *OOPS: 20 Life Lessons From the Fiascoes That Shaped America* (Collins, 2007).
Daily Mail (21 October 2010).

The metal cricket bat 金属板球拍

Martin Williamson, *Cricinfo Magazine* (25 September 2004).
Dennis Lillee, *Menace* (Hodder, 2003).
Australian Broadcasting Corporation (16 December 2003).

Bicycle polo & other lost Olympic sports 自行车马球及其他消失的奥运项目

Bicycle Polo Association of Great Britain, History & Rules (1938).
The Guardian (23 August 2004).
The Scotsman (16 August 2008).
The Observer (1 August 2004).

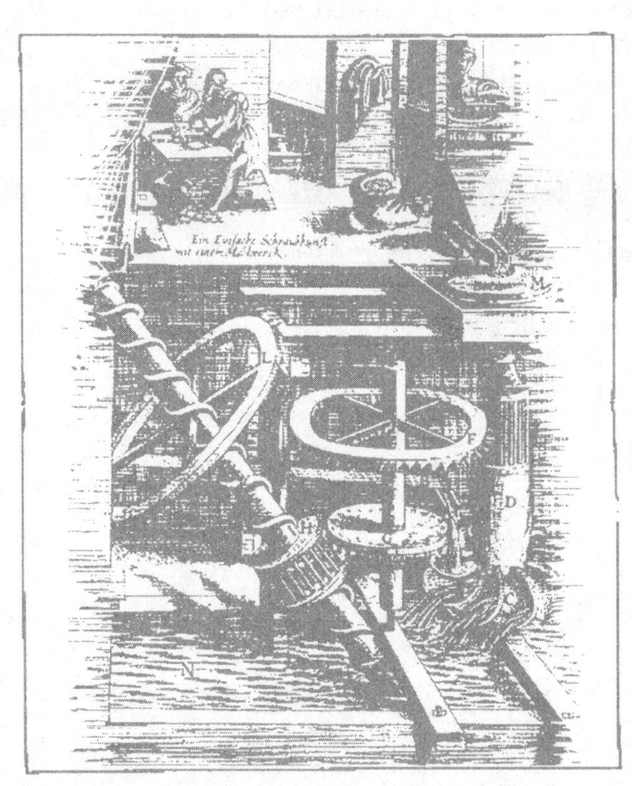

（下册完）